A Guide to...
Lories & Lorikeets
Their Management, Care & Breeding
Revised Edition

By Peter Odekerken

Published and Edited by ABK Publications ©

© **ABK Publications 2002**

**First Published 1995 by
ABK Publications
PO Box 6288,
South Tweed Heads,
NSW. 2486. Australia.
Revised Edition 2002**

ISBN 0 9577024 4 2

All rights reserved. No part of this publication may be reproduced, stored in any retrieval system, or transmitted in any form or by any means without the prior permission in writing of the publisher.

Front Cover:
Top left: Black-capped Lory
Bottom left: Varied Lorikeet
Bottom centre: Blue-fronted Rainbow Lorikeet
Bottom right: Duyvenbode's Lory
Back Cover: Yellow-bibbed Lory

All cover photographs by Peter Odekerken.

All other photographs by Peter Odekerken except where indicated.

Design, Type and Art: PrintHouse Multimedia
Colour Separations: Splash Colour
Printing: Southport Printing

CONTENTS

ABOUT THE AUTHOR — 5
DEDICATION — 5
ACKNOWLEDGEMENTS — 6
INTRODUCTION — 7
IN THE WILD — 9
   Taxonomy — 9
   Habitat Preference — 11
   Distribution — 11
   Interesting Facts — 12
   Breeding Habits – — 13
      Courtship, Incubation, Fledglings.
   Feeding Habits — 17
   Status and Conservation — 21
LORIES AND LORIKEETS IN CAPTIVITY — 25
   Selection — 26
   Transportation — 27
   Feeding — 28
   Housing — 31
   Breeding — 35
   Handrearing — 38
   Lories as Pets — 40
DISEASES AND DISORDERS — 42
   Crop Infection — 43
   Psittacosis (Chlamydophilosis) — 43
   Psittacine Circovirus (Psittacine Beak and Feather Disease PBFD) — 44
   Parasitic Diseases — 45
   Medication — 45
LORY AND LORIKEET SPECIES — 46
Distribution Map — 48
Chalcopsitta Genus — 50
   Black Lory — 51
   Duyvenbode's Lory — 54
   Yellow-streaked Lory — 57
   Cardinal Lory — 60
Eos Genus — 63
   Black-winged Lory — 64
   Blue-streaked Lory — 66
   Red Lory — 69
   Red and Blue Lory — 72
   Violet-necked Lory — 76
Pseudos Genus — 79
   Dusky Lory — 80

Trichoglossus Genus — 83
    Rainbow Lorikeet — 86
    Red-collared Lorikeet — 89
    Scaly-breasted Lorikeet — 91
    Ornate Lorikeet — 93
    Green-naped Lorikeet — 94
    Mitchell's Lorikeet — 95
    Weber's Lorikeet — 95
    Edward's Lorikeet — 96
    Rosenberg's Lorikeet — 96
    Massena's Lorikeet — 97
    Perfect Lorikeet — 97

Psitteuteles Genus — 98
    Goldie's Lorikeet — 99
    Mt Apo Lorikeet — 101
    Varied Lorikeet — 104

Lorius Genus — 108
    Black-capped Lory — 109
    Chattering Lory — 113
    Yellow-bibbed Lory — 117
    Purple-naped Lory — 120
    Purple-bellied Lory — 123

Charmosyna Genus — 126
    Stella's Lory — 127

Glossopsitta Genus — 130
    Musk Lorikeet — 131
    Little Lorikeet — 133
    Purple-crowned Lorikeet — 136

Neopsittacus Genus — 140
    Musschenbroek's Lorikeet — 141

MUTATIONS — 143
    Expectations for Basic Mutations — 144
    Primary Mutations – — 144
        Greygreen, Melanistic, Lutino
    Colour Combinations – — 145
        Melanistic Greygreen,
        Dilute Melanistic, Dilute Greygreen, Dilute Melanistic Greygreen,
        Other Combinations.

Table of Primary Mutations — 148
BIBLIOGRAPHY — 149
RECOMMENDED READING — 149
SPECIES NAME AND WEIGHT TABLE — 150

ABOUT THE AUTHOR

Photograph of author courtesy of Anne Love.

Peter Odekerken has always had a love for wildlife, particularly birds. He started keeping finches when he was 11 years old and eventually progressed to his first parrots, namely Bourke's Parrakeets not long after. Over a 39 year period he has bred many of the Australian species of parrots and became enraptured by the beauty and character of lorikeets. Peter lived in South Africa for 13 years and during this time kept many non-Australian species of lories and lorikeets which he successfully bred. He was the first person in South Africa to breed the Purple-naped Lory and continued to do so during the mid 1970s. He was awarded a silver medal by the Parrot Society of UK for this breeding as well as the award for a first breeding in South Africa by the South African National Cage Bird Association.

He then returned to his native Australia and now runs a company which installs polyethylene pipe in Australia, the South Pacific and South-East Asia. When he gets the chance he takes every opportunity to study birds in the wild and has been fortunate to have seen many species of lories and lorikeets in their natural habitat. Peter decided to take up photographing parrots because there were very few well-illustrated books on parrots in the 1970s and he wanted a collection of photographs for his own benefit and for use at his numerous lectures. His collection of photographs was taken all over the world at avicultural friends' aviaries, bird parks and zoos. Peter has lectured throughout Australia, New Zealand, Holland and South Africa, and enjoys increasing his knowledge by speaking to fellow aviculturists.

Peter records digital video and images which he incorporates into his lecture presentations. As a dedicated Australian aviculturist, Peter hopes for a better understanding between government organisations and the avicultural community due to the very real need to conserve all wildlife. 'The future for avian wildlife is bleak with the continual pressure exerted by most notably, habitat destruction and of course in some cases capture for aviculture', says Peter. 'It must be remembered and recognised that aviculture can be of great assistance in the conservation of the threatened global birdlife'. Peter also looks forward to speaking at avicultural conventions throughout the world on his favourite topic, lories and lorikeets.

DEDICATION

For many of us, the fascination of birds originally started our interest because we wondered at their ability to fly. I was also drawn to them because they are particularly beautiful in my native Australia. Many species of parrot frequented my garden and I just could not get over their colours.

Brisbane has two commonly seen species of lorikeet, namely the Scaly-breasted Lorikeet *Trichoglossus chlorolepidotus* and the Rainbow Lorikeet *Trichoglossus haematodus moluccanus*, both resident in our gardens for most of the year. I suppose it is to these birds that my long suffering wife and daughter can lay blame. They have often felt that they came second in the scheme of things. It is to them, Kathy and Tracie that I dedicate this book.

ACKNOWLEDGEMENTS

Obviously through the years I have met many people who have contributed to this book, providing information and friendship. I would hate to leave out anybody but expect I will and it will be regretted if I do so.

Warren Allen – Northern Territory, Australia.
David Austin – Queensland, Australia.
Thomas and Angela Arndt – Germany.
Fred Barnicoat – South Africa.
Peter Blom – Netherlands.
Dr Danny Brown BVSc (Hons) – Queensland, Australia.
Armin and Karen Brockner – Germany.
Trevor, Maura and Mark Buckell – England.
Peter and Dianne Clark – Papua New Guinea.
Ron and Dorothy Cleaver – Queensland, Australia.
Brian Coates – Queensland, Australia.
Neville and Enid Connors – New South Wales, Australia.
Peter Coyle – New Zealand.
Antonio de Dios, 'Birds International' – Philippines.
Gordon and Julie Dosser – Victoria, Australia.
Damian and Sheryl Dunemann – Queensland, Australia.
Wim Engelaer – Netherlands.
Frans-Josef Engels – Germany.
Dr. Louie Fillipich – Queensland, Australia.
Bill and Muriel Fulham – Queensland, Australia.
Dr. Adrian Gallagher – Queensland, Australia.
Mike Gammond – Los Palmitos Park, Canary Islands.
Will Glynn – Papua New Guinea Bird Society.
Jim and Wendy Hill – Queensland, Australia.
Stephen and Leanne Hill – Queensland, Australia.
Jos and Mariette Hubers – Netherlands.
Wolfgang Kiessling – Loro Parque, Canary Islands.
Lawrence Kuah – Singapore.
Rosemary Low – England.
Anne and Beresford Love – Australia.
Bob and Helen Lyle – Northern Territory, Australia.
Graham Matthews – South Australia, Australia
Dr Terry Martin BVSc – Queensland, Australia.
Russell McAllister – New South Wales, Australia.
Hank and Jan Michoruis – Netherlands.
Eric Mickeler – France.
Maureen, Chris and the late Richard Muhvich – New South Wales, Australia.
Bettina and Harold Muller – Germany.
Klaus Paulmann, Los Palmitos Bird Park – Canary Islands.
The late Bill Peckover – Queensland, Australia.
Thomas Petzen – Germany.
The late George and Marion Pollock – New Zealand.
The late Fred Shaw Mayer – Queensland, Australia.
Peter Slater – Queensland, Australia.
Phil Smith – Victoria, Australia.
Des Spittall – Queensland, Australia.
Ron and Dawn Stewart – New Zealand.
Patrick Tay – Singapore.
Paul Tiskens – Germany.
Yvan Vaes – Belgium.
Gert Van Dooren – Netherlands.
Tony and Jude Vaughan – Queensland, Australia.
Hans Visser – Netherlands.
Rob and June West – New Zealand.
Shaun Wilkinson – South Africa.
Gavin and Kirsteen Zeitsman – South Africa.
Johnny and Thora Zietsman – South Africa.

I would like to thank the following people for their photographic contributions:
A Brockner, Dr M Cannon, W Caldwell, R Dunn, J Kenning, V Nicholson, C Slaney, B Willis, S Wilson and A Zarra.

Finally, I would like to express my gratitude to Nigel and Sheryll Steele-Boyce of **ABK Publications** whose guidance, patience and persistence ultimately prevailed in the publication of the original book and this revised edition.

INTRODUCTION

Lories and lorikeets belong to the parrot family and have the typical physical attributes which make the parrots so well known. Even people not interested in birds can describe the shape of a parrot.

For the sake of convenience I will generally use the term lory or lorikeet in the text instead of repeating myself using both terms.

The name lorikeet can be roughly used to describe those species with a long tapered tail, whereas the name lory is used for species with short rounded tails. This can be equated with the terminology, parrot and parrakeet. Invariably most members of the sub-family *Loriinae*, to which lories and lorikeets are ascribed, are extremely gaudy. Reds, greens and blues predominate but the entire spectrum of the rainbow can be found and some species can even boast most of the colours within their own attire.

I have often wondered whether these colourful birds would be vulnerable to attack from their predators, but believe me they are remarkably well camouflaged in the wild. I have been fortunate enough to have viewed many species in the wild, including such colourful forms as the Black-capped Lory *Lorius lory lory* that can prove very difficult to spot even with binoculars, when perched in the tree tops. Their voice may indicate their presence, but often the only thing that gives them away is their movement from branch to branch.

We in Australia are fortunate to have lorikeets visit our gardens in many of our capital cities. Therefore most Australians are familiar with these birds, whereas people living outside their distribution can only see them in zoos or avicultural collections.

Whiskered Lorikeet – the only parrot with 14 tail feathers. All other parrots have 12 tail feathers.

Although I have observed these birds in my garden for many years I still never fail to find interest in them as they clamber in a flowering grevillea outside my bedroom window.

I must admit that in trying to obtain information on the group from publications, it was drawn to my attention, how little we know of these birds. Nesting habits are often undescribed and very little detail is given on the majority of species. This is largely due to their remote locations but also to a lack of enthusiasm by observers to communicate their observations. When reading through literature it is unfortunate that most records of bird species in the wild relate to sightings in a particular area.

Also, many observations are recorded on a personal basis and are never made available to other enthusiasts in periodicals. We can chastise both ornithologists and aviculturists for not making this information available. I realise that it is extremely difficult as often very little behaviour occurs during a period of observation in the wild. However, many articles in ornithological publications are of little use in understanding bird species. At most they provide a record of distribution and status in an area at a certain time of a particular year. Aviculturists write articles about the breeding and certainly the keeping of lories and lorikeets in captivity. Many do not even bother recording incubation periods, descriptions of their young, rearing periods, behaviour and the list goes on and on. Aviculturists have the rare opportunity to describe the behaviour of species but this is seldom recorded. We have at our beck and call species which have never had simple observations such as courtship display recorded.

I hope in reading this book that more people will put pen to paper to add to our knowledge. This book is far from being the final chapter in our knowledge about this group of birds, in fact we are only at the beginning.

Knowing that many species are candidates for extinction, I can only hope that it will convince those people with access to information to make it available to other interested parties.

IN THE WILD

Taxonomy

There are 12 genera, 55 species and approximately 63 subspecies of lories and lorikeets existing in the world today. This depends on whether the New Britain or Stresemann's Lory *Lorius amabilis* and the Blue-thighed Lory *L. tibialis* can be considered as species. I say approximately, because subspecies validity is a debatable point especially when investigating the Rainbow Lorikeet group *Trichoglossus haematodus*.

I am not intending to go into any detail on taxonomy, but I would like to express my views on some classifications. These are not supported by scientific qualifications, but are based on my practical experience with the groups over the years.

Simon Joshua (*Lori Journaal Internationaal* Volume 1994 No.1) from the UK has been researching the taxa based on DNA profiles and many interesting views will be expressed when his research is eventually published.

For many years I have noticed the close relationship of the Dusky Lory, genus *Pseudeos* with the genus *Chalcopsitta*. Many scientific analysts have not had the good fortune to be able to study their taxa in the wild or in captivity as far as behavioural aspects are concerned. I believe that behaviour plays an important role in taxonomy and supports the physical assessment of taxa.

Although the Dusky Lory is naturally smaller than *Chalcopsitta*, there are some important physical characteristics relevant in both genera. The bare skin around the eye and lower mandible and the similarity in colouration to the Duyvenbode's Lory *Chalcopsitta duivenbodei duivenbodei* are indicative of a close relationship. This relationship is however supported by behavioural characteristics. Especially interesting is the display of ruffling the head, body and rump feathers with tail

Rajah Lory hen.

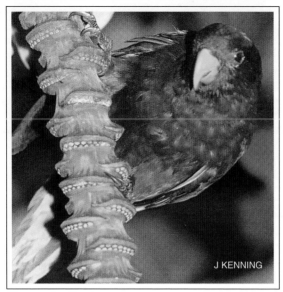

Ponape Lorikeet

fanned and wings draped loosely at the side of the body. The call and screeching of *Pseudeos* are less noisy than *Chalcopsitta*, but otherwise very similar. Begging for food by immatures is alike for posture and voice. I was pleased to see in an article published in *Lori Journaal Internationaal* (Volume 1994 No.1) in which Joshua noted a close relationship between both genera based on DNA profiling.

Lorius is also an interesting group which I would never dispute without more scientific support particularly in relation to DNA profiles. Again within this genus I feel that the group can be analysed and separated into the Chattering and Black-capped *Lorius* group with the Yellow-bibbed Lory *Lorius chlorocercus* showing a tendency towards the Collared Lory *Phigys solitarius*. I would propose to subdivide and create new genera for these species because of the physical and behavioural characteristics that define them separately. I have not had the opportunity to evaluate the White-naped Lory *Lorius albidinuchus*, however, it is more than likely linked between the large Black-capped group and the Yellow-bibbed Lory.

The Yellow-bibbed Lory is a slimmer build and smaller size than the New Guinea Black-capped Lory *Lorius lory lory* and is certainly a much more active bird in captivity. I certainly feel there is a link which would deserve research between it and the Collared Lory. The Collared Lory is also an active bird having a purple-black cap which is noticeable on the Yellow-bibbed Lory in strong sunlight. In fact, the black crescents at each end of the yellow bib on *L. chlorocercus* are actually dark purple. There is also no doubt that the Collared Lory links to the genus *Vini*, but I feel that *Charmosyna* is not as closely related to *Vini* as has sometimes been expressed. Please remember that the entire family is extremely closely related.

The Ponape Lorikeet *Trichoglossus rubiginosus* has been considered by some to be associated with the Cardinal Lory *Chalcopsitta cardinalis*. However, I have observed the Ponape Lorikeet both in the wild and in captivity and conclude that it is not closely related to the Cardinal Lory.

An extremely interesting genus is *Psitteuleles* which I am not prepared to alter until more DNA profiling takes place. I am glad that Birds Australia has kept the Varied Lorikeet in this genus after Forshaw and others had altered it to *Trichoglossus*. Forshaw never

Rajah Lory cock showing an unusual amount of red colouring.

did explain why he considered it to be *Trichoglossus* and unfortunately many people blindly followed suit with no justification.

It is my opinion that *Psitteuleles* could possibly be split because there are certainly close physical characteristics between the Varied Lorikeet *P. versicolor* and the Goldie's Lorikeet *P. goldiei*. The second group would be the Mt. Apo *Trichoglossus johnstoniae*, the Meyer's *T. flavoviridis meyeri*, the Yellow and Green *T. f. flavoviridis* and the Iris *T. iris iris* Lorikeets . Perhaps the groups should never be separated as I see some measure of similarity between the orange-coloured eye ring of the Yellow and Green Lorikeet and the white eye ring of the Varied Lorikeet. The Varied Lorikeet certainly exhibits courtship behaviour which is totally alien to the *Trichoglossus* genus. Both adults will perch next to each other, stretch and sway or arch away from each other in unison making an indiscernible hiss in the process. This is not found in the *Trichoglossus* group that have their own similar courtship behaviour. It may seem insignificant to taxonomists but I cannot ignore the abdomen streaks and the red cap in the Varied and Goldie's Lorikeets and in my second group, the body scallops. Perhaps the Iris Lorikeet could be considered a link between these groups as it does exhibit red on the forehead and some form of scalloping.

I have to admit that even DNA profiling may never totally refute the arguments that taxonomists have had in the past and will no doubt have in the future.

Habitat Preference

Lories and lorikeets are found from one extreme to the other in regard to the habitat they have successfully adapted to.

The Varied Lorikeet is found in Australia's semi-desert terrain around Mt Isa, northern central Queensland and the Northern Territory where daytime temperatures are extremely high, humidity low and night-time temperatures can reach below freezing in winter.

Mt Isa, Australia – Varied Lorikeet habitat.

Many New Guinea species flourish in lowland and montane rainforest and other species have adapted to colder climates such as the alpine tracts on the slopes of the high peaks of the central backbone of New Guinea. Indeed it is possible to see the Whiskered Lorikeet *Oreopsittacus arfaki arfaki* feeding with a snow-capped mountain as a backdrop. They have also shown remarkable resilience in adapting to disturbed habitats where indigenous trees have been replaced by exotics. Dusky Lories *Pseudeos fuscata* in their hundreds can be found west of Port Moresby, Papua New Guinea in coconut plantations. The Purple-bellied Lory *Lorius hypoinochrous hypoinochrous* has taken to coconut plantations to such a degree that it is considered a pest in the Rabaul region.

Many other introduced species of trees and shrubs which have been planted in parks and gardens have also become a food source.

Distribution

The distribution of lories and lorikeets radiate from a nucleus known as New Guinea. From there they radiate west to the tourist paradise of Bali, Indonesia and are found throughout the Lesser Sunda Islands to the Moluccas and as far north as Mindanao in the Philippines and Pohnpei in Micronesia. They radiate east to the Marquesas Islands which lie between Tahiti and Hawaii, then further south to Henderson Island just near

Coconut Palms – Dusky Lorikeet habitat.

the island of Pitcairn, famous for the story of 'Mutiny on the Bounty'. They are also found throughout Australia and as far south as Tasmania.

This uniform family of parrots therefore inhabits some of the most remote places on earth. The family has never been considered to possess particularly great flying strength and yet they have colonised much of the greatest expanse of water on earth, the Pacific Ocean.

Birds of the genus *Trichoglossus*, also referred to as the Rainbow Lorikeet group, have proved extremely interesting for taxonomists because of the various forms found within the species. The species has the ability to cover vast distances over sea and land and has strong flight capabilities which no doubt helped it to colonise such a large region.

However, one must marvel at the members of the genus *Vini*, that have distributed themselves onto many small islands within the vast ocean regions. They are certainly not very robust little birds and their flight capabilities are not in the same league as the *Trichoglossus* genus.

Interesting Facts

Naturally the age a bird attains in the wild is impossible to establish. However a Rainbow Lorikeet, banded on 5 September 1964, was recovered on 11 October 1983. This information was published by 'The Australian Bird and Bat Banding Scheme' in February 1987. The bird was more than 19 years old. I feel that this is exceptional in the wild but is certainly possible in captivity.

Of interest also is that banding records show that the greatest distance travelled from one banding site to another is as follows:

SPECIES	DISTANCE	NUMBER OF RECOVERIES
Rainbow Lorikeet	80km	39
Scaly-breasted Lorikeet	60km	4
Musk Lorikeet	59km	4
Purple-crowned Lorikeet	53km	13

Unfortunately, records are in short supply, but at least it gives us some idea of nomadism and longevity.

Purple-crowned Lorikeet cock.

Breeding Habits

When conditions are favourable, species breed in hollows found in trees. There are records of the Olive-green Lorikeet *Trichoglossus haematodus flavicans* breeding in underground burrows which were actually dug by the lorikeets. According to the late Bill Peckover (pers. comm.) while on the islet of Poy-yai, part of the Admiralty group, north of the Papua New Guinea mainland, there seemed to be alternative nest sites in trees, but *T. h. flavicans* preferred to nest and roost on the ground.

Family of Musk Lorikeets from right: two fledglings, hen and cock.

These observations were confirmed by the local residents. Species of the *Vini* genus are said to breed inside old damaged coconuts and the leaf oxits in the coconut crown. It seems however that the last two examples are yet to be verified, and I must admit that it would certainly be unusual to find the birds breeding in coconuts. Coconuts are extremely difficult to break into and would drop to the ground in most cases before decomposing sufficiently for the lories to break through and gain access.

Small species such as those found in the *Charmosyna* genus are known to excavate their nest entrance and chamber in the base of arboreal plants such as tree ferns, epiphytes and other plants that create a fibrous root ball. The birds cling busily to this fibrous base and industriously excavate the entrance and then the chamber, which when complete provides a base to lay their eggs on, not unlike the texture of peat moss. During August 1995, I found nests such as this on Lihir Island, New Ireland Province, Papua New Guinea. Unfortunately, the Red-flanked Lorikeets *C. placentis placentis* which had been so hard working, were observed by the local young children who captured them at night in their new home and then promptly ate them.

Generally lories prefer hollow limbs or holes in the trunk of a tree. No lining is used except for feathers plucked from the brood patch. The eggs are usually laid on top of decayed wood that has accumulated over a period or has been chewed from the sides of the nest by its occupants.

Most species lay two white oval-shaped eggs but there are exceptions. The Perfect Lorikeet *Trichoglossus euteles* regularly lays three eggs, and Australia's smaller *Glossopsitta* species and the Varied Lorikeet *Psitteuteles versicolor* can lay up to five eggs, with three to four being the norm. Eggs are generally laid on alternate days or longer.

Courtship

Courtship displays have many affinities within each species. Typically courtship involves head bobbing, stretching, jumping on the spot and pupil dilation. 'Hissing' is best described as a form of vocalisation which seems more noticeable during courtship than at other times. The birds open their bills and blow air out of their lungs with the tongue usually protruding from the bill. In some of the smaller species the sound made can be quite faint, whereas larger species can be heard from quite some distance.

A typical display of the *Trichoglossus* genus would be screeching, hissing, wing-whirring and pupil dilation as preliminary behaviour. The pair will then sit closer to one another and both partners will stretch and arch towards each other with the cock becoming more dominant. The hen will then observe the cock's display and does not perform any other movement. The cock will begin to bob his head and dilate his pupils, then stretch and arch over the hen. If she is ready to mate, she will respond by crouching low on the perch with wings held at her side lifting the bend of her wings away from the body to afford a platform of support. The rump is then pushed up as the highest point of the body. She will only accept the cock if she is ready to mate and it may take a fairly elaborate performance before she goes into her submissive posture. The cock then mounts the hen and grips her with both feet so that she has to support his weight. The cock moves his vent against hers pushing her tail aside. He moves from side to side occasionally biting softly at her plumage or bend of wing. Copulation can take some time and on orgasm the cock arches backward and often flaps his wings. He then quietly dismounts and the pair will often mutually preen afterwards.

Naturally the wing-whirring is pronounced in species with bold colours under their wings but many of the smaller species do not wing-whirr. As an example I have never observed the *Glossopsitta* species or the Varied Lorikeet wing-whirr. The courtship of the Little Lorikeet *Glossopsitta pusilla* is quite different from *Trichoglossus*. The cock has his chest feathers puffed up whilst he bobs close to the hen emitting a low grating sound. He pushes himself against her to force her into a submissive posture. His head feathers can also be ruffled and sometimes he fans out his tail in the display. The tail has much red underneath and he therefore uses this to advantage in impressing the hen. A Little Lorikeet has a rather plain underwing colour and this could not be used to advantage when wing-whirring. The Varied Lorikeet also does not wing-whirr for similar reasons. The cock and hen Varied Lorikeet sit side by side with the cock bobbing up and down emitting an unusual 'chirrup' type call. Some sign seems to be given to the hen because the cock will stretch and arch away from her which she does in unison at the command. This performance will be repeated until she accepts him to mate. His bright red cap and the upper chest are both fluffed out during the display.

Striated Lorikeets *Charmosyna multistriata* of central New Guinea have a display which again does not involve wing-whirring. The cock fluffs up his breast feathers highlighting the streaks to impress the hen. He accompanies this with arching and bowing over the hen. This makes her crouch on the perch and he will then attempt to mount her. If she is not ready to accept his intentions she moves away and he will follow her to start all over again until she no doubt accepts his advances. In all probability he 'hisses' and warbles to her during his advances, but I was unable to confirm this at my sightings due to ambient background noise.

Fairy Lorikeets *Charmosyna pulchella pulchella* of the same region also seem to perform a similar display. The cock's yellow streaks are very noticeable against the predominant red breast feathers.

Yellow-bibbed Lories *Lorius chlorocercus* from the Solomon Islands have a striking underwing pattern but they do not wing-whirr as do the *Trichoglossus* genus. Cocks and hens can be rather aggressive in their behaviour, with birds actually rolling on their backs and using the beak and legs to wrestle each other. Wings are held open during these bouts, but once the cock shows his intentions he will keep his wings in place along his sides. He then bobs his head up and down and stretches his legs and body while perched next to the hen. When she crouches on the perch, he will attempt copulation. Mutual preening often occurs before the display and will occur after copulation or the pair will sit next to each other preening themselves.

Chattering Lories *Lorius garrulus garrulus* have a display which predominantly involves head bobbing, body stretching and 'hissing'. The hen is rather submissive or is intent in paying attention to the cock. The larger lories have a loud hiss and in the case of Chattering Lories they can increase the intensity and make a gruff sounding hiss when extremely excited.

Pupil dilation has not been mentioned in the displays but is noticeable in most species. It is particularly apparent when the pupil is surrounded by a red or yellow iris. As the pupil contracts, the iris naturally has a larger area and the red or yellow is highlighted. Lories are not the only parrots that dilate the pupils in display. Pupil dilation may also occur during extreme excitement, but is certainly most prominent during sexual arousal.

It would be particularly satisfying if more authors would write about their observations so that we can compare the display between species.

Incubation

Incubation periods vary depending on the particular species and usually start after the first egg is laid. It has been recorded that *Charmosyna* and *Vini* species can wait a few days before incubation commences. Cocks will also incubate the clutch, particularly during the day. Incubation periods vary according to ambient conditions. Obviously high temperatures during the incubation period can assist in the egg hatching, resulting in a shorter term than usually expected. The converse is true in colder weather. The incubation period can vary from 20–28 days depending on the species.

Fledglings

Young can be heard peeping inside the shell close to hatching. Most species are covered by a fine, white wispy down and have pink skin. The eyes are closed and usually open fully at approximately ten days of age, again depending on the species. During the first ten days at least, the hen sits tightly with her young, keeping them warm or comforting them if it is hot enough not to brood them. During the first day or two the young are generally fed on their backs as they are very unstable on their legs. Once they gain strength, they will push strongly for food using the abdomen for support until their legs are strong enough to support them. After a few days, again depending on the species, dark pin-feathers can be seen under the translucent skin. This secondary down, usually a grey colour, provides greater warmth for the young bird in preparation for the hen to leave the nest on a regular basis to help the cock search for food. In the Musk Lorikeet *Glossopsitta concinna* I have found this grey down to be very thick and wool-like. I assume this is due to the colder climatic conditions found in its habitat. Rainbow

Breeding pair of Purple-crowned Lorikeets with four young.

A recently fledged Red and Blue Lory.

Lorikeets can also show heavy down particularly when breeding during the colder months. It is my opinion that the thickness of this down is regulated by the prevailing climatic conditions at the time. I cannot substantiate this as I have never been in the position to accumulate enough data to confirm it.

On the other end of the scale, Varied Lorikeets, which are a tropical species, do not attain this secondary down and the young remain naked until their mature feathers erupt from their sheaths. For this reason, Varied Lorikeets would be unlikely to breed in desert areas such as the Mt Isa district of northern central Queensland in winter, because the young would perish. In captivity, the Varied Lorikeet hen does not brood her young after about ten days, even though she may sleep with the cock in the nest every night.

When breeding Varied Lorikeets in colder climates the crucial time for young is between the age of approximately ten days and feathering. Young need a nestbox insulated against the cold or additional heating. Other aviculturists may dispute this because their pairs are better brooders. However, this has been my experience over the years with a number of pairs being kept in the Brisbane region of Queensland. Varied Lorikeets breed during the winter months and cold nights can be experienced in southern Australia.

In captivity the nestbox can become fouled by the liquid excreta of the young. It seems that this is not a problem when breeding occurs in the wild as the base of the nest is composed of decayed wood which could be many centimetres deep. I have observed in sawmills hollow trunks with a decomposed central area of wood running the entire length, often metres long. This area close to the base of a nest hollow would absorb the liquid content of the excreta. I have found that if the nest is not kept clean, deaths can occur in young if they become wet and cold due to accumulated excreta. This would occasionally occur in the wild where the decomposed wood base of a hollow did not have sufficient depth.

A dark hollow must be a difficult place for a bird to feed its young. This becomes an important factor in feeding as the parents must locate the chick's bill and then regurgitate the food. It is my opinion that some species have adapted in such a way to assist in this feeding procedure. The genus *Chalcopsitta* have naked dark grey or black skin around the base of the lower mandible and around the eye, known as the periorbital ring. In this genus the young have white skin in these two areas which I believe has been adapted to assist the parent bird in locating the head and the bill of the young to facilitate feeding. On leaving the nest this white skin very quickly attains its adult colouration.

In the genus *Lorius* I have noted that the Purple-naped Lory *L. domicellus* has, on the lower mandible, a fleshy bulbous ridge which runs from the base to approximately halfway along the mandible. This fleshy area, which is white in colour, contrasts with the black on the rest of the bill. On the upper mandible, there is a corresponding area, which is slightly smaller. The lower mandible is noticeably wider than the upper mandible at this stage. The lower mandible probably acts as a scoop for feeding or more correctly a larger area to fill. This white fleshy region is sensitive to touch and the youngster, particularly when still blind, reacts to its touch. No doubt when the parent's bill comes in contact with this sensitive area the youngster jerks forward in anticipation of food. I have touched this area on a youngster which was being handreared and the response was immediate begging behaviour. Black-capped Lories have the same white area which must assist the parents in locating chicks in the dark nest for feeding.

Depending on the species, young leave the nest from between six weeks to three months of age. For the size of the bird, this is an incredibly long period in the nest. In the wild it would be unusual in the large species for a pair to breed more than once a year and I do not believe it would be very common for the smaller species to breed twice a year.

In most cases, the young leave the nest as duller versions of their parents. If the adult bill colour is red, then the young invariably can be distinguished by their dark brown to black bills. These gradually change into the adult colour and the immature plumage is usually moulted out in the first six months.

Most young lories and lorikeets are steady and extremely capable, soon learning from their parents to forage for themselves.

As in all species of animals, the mortality rate is highest in young birds and a very high percentage do not make it through the first year. This is understandable considering the factors applicable when a stable population in an area breeds. Obviously a particular region would have a stable population of a certain number of pairs. This varies depending on the availability of food and nesting sites in the area. If this area has a good season the usual number of pairs may swell for a season or number of seasons, because more young reached maturity due to an ample food supply. However, a breeding nucleus has the ability of doubling its numbers in a breeding season. Once the breeding has ceased, seasonal conditions will reduce the population numbers similar to that previously sustained by the region. Young are incapable of commanding a food source from aggressive adults and can starve to death. Also, they are still very vulnerable to predation whereas an adult has already experienced attacks and generally has a better chance of survival.

Sexual maturity is attained at about six months of age in some small species. For example a pair of my handreared Purple-crowned Lorikeets *Glossopsitta porphyrocephala* bred and were successful in raising one youngster under captive conditions. In the wild, I doubt that breeding would occur before the next season, at least 12 months away. Larger species such as *Chalcopsitta*, may take up to four years to breed but have been known to breed much earlier in captivity.

Feeding Habits

To harvest their food in the wild, lories and lorikeets possess a very special adaption, papillae on the end of the tongue. Papillae can be retracted from their extended form to be folded back in a protected position. In this position the tongue looks similar to that of most other forms of parrot – like birds, that is, large and fleshy. The papillae must be a very sensitive organ otherwise there would be no need for this protection. When watching lorikeets feeding on pollen, it will be noticed

Yellow-backed Chattering Lory showing papillae on the tongue.

just how well developed and efficient this collecting organ is. It reminds me of a cat licking milk. The birds harvest pollen quickly moving from flower to flower thereby

Wild Rainbow and Scaly-breasted Lorikeets being fed in the author's garden.

performing the all important function of pollination. Lorikeets can be seen with pollen adhering to their foreheads where the stamens have touched the feathers. They feed not only on pollen but also consume large quantities of nectar. This has been discussed at length by Churchill and Christensen (1970) whose observations of the Purple-crowned Lorikeet of southern Australia confirmed that this species subsisted almost entirely on pollen at certain times of the year. However, I feel that some species have adapted to utilise various food sources more so than others. For instance, the Musschenbroek's Lorikeet *Neopsittacus musschenbroekii* has a much stouter bill than most similar sized lorikeets, and possibly utilises seeds or some properties in bark similar to that recorded for Fig and Pygmy Parrots.

The birds have also needed to adapt themselves to reach blossoms and fruit. To this end they have powerful feet capable of gripping slender twigs where flowers and their fruits are usually supported. They also have short legs, more particularly referred to anatomically as the tarsus of the leg. This surely assists the birds to clamber skillfully through the foliage. Obviously long legs such as those of a stork, would make it difficult to move over and between blossoms, leaves and twigs. The feet are capable of gripping leaves to steady the bird in position to obtain sustenance.

The beak is of typical parrot shape ie hook-billed, but is not as solidly constructed as in seed-eating parrots. The function of the beak in sourcing food is merely to access the inside of fruits and make it available to the tongue's papillae. The internal pulp is usually discarded once the juice has been extracted. Lories and lorikeets manipulate the pulp between their mandibles and the tongue licks off the juice, then with a quick shake of the head, the used pulp is discarded. The beak easily tears the skin of an incautious aviculturist and proves a formidable weapon when used in defence or attack. Fortunately lories and lorikeets very seldom persist with their attack and it is unusual for flock members to kill or maim each other. However, on one occasion I witnessed a pair of Scaly-breasted Lorikeets *Trichoglossus chlorolepidotus* defending their nest site from another pair. During this fight two birds interlocked and came tumbling to the ground. Both partners were locked together and continued their fight on the ground oblivious to the numerous school children walking near them. The birds eventually released each other on the ground and flew back to the nest site, screeching all the while. Almost immediately they again flew at each other and again came tumbling to the ground. However, on the second occasion, on release, one of the birds could not fly. Whether the damaged wing had been directly inflicted with the beak or indirectly damaged by hitting the ground is

Wild Rainbow Lorikeet feeding in a grevillea bush.

Pair of Varied Lorikeets feeding on gum blossoms.

not known. This does illustrate that damage can occur, resulting in death, for I am sure that the unfortunate bird would have fallen prey to a predator.

If a pair of lorikeets decide to attack a much larger opponent, they work to disadvantage the bird by placing themselves on either side of their opponent. One partner will distract the interloper while the other member of the pair will lunge and bite the unsuspecting bird. I have heard of many fatalities in aviculture because the owners were unaware of the aggressive nature of lorikeets and did not believe that a small pair of Rainbow Lorikeets could kill a much larger cockatoo.

Lories and lorikeets are generally flocking species that tend to congregate at food trees during the breeding season. They will collect food at these locations returning to the nest to feed young or the sitting hen. Therefore, even when pairs separate from the flock to breed privately, there is still much contact with other members of the species. During these feeding sessions, lorikeets can be heard squabbling with rivals over a food source. The disturbances are rarely serious and the most aggressive display intimidates the opposition to give way and no further action takes place.

Some of the larger species such as the genus *Lorius* tend to remain in family groups. These are usually pairs and their young of the season seen gathering food together. On occasion these family groups will congregate with others in the same tree, if a particularly abundant supply of blossom or fruit is available.

It is noticeable to see social interaction between family groups but they still tend to retain autonomy. Naturally, as young become more independent they may form a close association with another bird of a different family. It is presumed that they will become bonded and establish themselves in their own niche as a breeding pair.

It is not unknown for lories to take insects, in some cases incidentally while licking pollen and nectar, and there are records of species being observed eating insect larvae. Records are so scarce, or possibly and more correctly, seldom identified during observation that we can only assume that insects are taken to provide protein in their diet. Stomach content analysis has identified intake of insects, but again so little information is at hand that it is difficult or certainly unwarranted to assume that all species gather insects.

In my travels, particularly in Australia, I have observed flowering eucalyptus in some regions and at certain times of the year where there are no lorikeets present. The same species of eucalyptus may be in flower in another region and hoards of lorikeets, honeyeaters and insects such as bees will be present. Even though trees seem to be capable of providing food because they are flowering, this does not suggest that they are actually producing the right food for lorikeets at that time. In fact, I believe that lories gain much of their food intake from nectar as opposed to pollen which is contrary to some information. The simple reason is that I have found that the lorikeets display limited interest in the flowers of eucalyptus and grevillea, even though they possess pollen, until the nectar is present at the base of the stamens. Often this period is referred to as the 'nectar flow'. Once the flowers produce this flow there is an unbelievable amount of activity with nectar-eating birds, mammals and insects continually visiting the source day and night. When walking around flowering woodland, you can even smell the nectar flow from eucalyptus. It is only during this time that large congregations of lorikeets will be present. Grevilleas in my garden have flowers for a great part of the year with very little bird activity until the nectar globules form at the base of the flower. In fact, the nectar flow can be so copious that when shaking a flowering branch you can be covered by nectar. When I cut these flowers for my captive lories and lorikeets I take great care to avoid displacing the nectar globules before placing them in with the birds.

For these reasons lories need to cover reasonable distances to obtain food, particularly when breeding. Naturally a nest hollow may not be in close proximity to flowering trees. Also, many species outside the breeding season form large flocks, probably for protection from aerial predators such as hawks and falcons. These large flocks often have prescribed roosting trees which can be a considerable distance from the food source. In many island species this may even mean crossing substantial distances of the ocean.

Lories are generally nomadically inclined because they need to follow the availability of flowering trees. They have been known to fly quite long distances between their feeding grounds and roosting sites.

These roosting sites can have thousands of birds arrive at dusk usually accompanied by much shrieking. The noise can be overwhelming and flocks in suburban gardens are often the cause of many complaints by the local residents.

In Townsville, North Queensland, a large roosting flock congregated outside the Travelodge Hotel in trees planted in the mall of the central business district. The birds continued all night due to the street lighting. Naturally, flocks tend to settle down during the night where they are not disturbed by lights. During the dark hours, particularly on moonlit nights, bickering can be heard between birds.

When coming in to roost, individuals and pairs will squabble for position. Usually the mornings are more orderly with pairs mutually preening and flying off to feed as the light of day increases.

On witnessing lorikeets feeding in blossom, I believe that large congregations would be far more effective in avoiding predators. Lorikeets tend to bury their heads in flowers to get to the nectar and are very vulnerable to attack at this time, particularly by goshawks. In a large group there will always be some members on the lookout to raise the alarm while others are occupied feeding.

Although nectar and fruit obviously provide a certain amount of moisture required by lories, they still drink water, particularly in dry areas on hot days. Lories will lick water from wet leaves after rain, and I have observed these arboreal birds hop to the ground to reach water at dams and urban sprinklers. Most of their bathing is done amongst wet leaves during or after rain and with much gusto. They fluff their body feathers, flap their wings and rub themselves amongst the leaves, all the while screeching in what I assume is ecstasy.

I have seen Red-collared Lorikeets *Trichoglossus haematodus rubritorquis* bathe on the grass under sprinklers on a median strip right next to the Stuart Highway at Adelaide River, Northern Territory. This indicates that lorikeets will come to the ground in dry periods for various reasons but it must be stated that this is unusual. Lorikeets that come to my garden for a free feed during a poor season will also readily hop on the ground in search of food scattered by other birds.

A trio of Tahiti Blue Lories which are extremely rare in the wild and captivity.

Status and Conservation

Generally lories and lorikeets are considered to be common species and only some island species are vulnerable at present. We must not lose sight of the fact that habitat destruction far outweighs any other factor in promoting the extinction of a species. In Australia, aviculture has been attacked by 'greenies', animal liberation groups, the scientific community as well as government authorities and most state wildlife agencies. This is not peculiar to Australia, but is a worldwide problem for aviculture. Unfortunately, the fact remains that if we are to consider our obligation as human beings to preserve wildlife on our planet, we need to work together, and aviculture must be truly recognised.

I know of no aviculturist who would wish to see all parrots become extinct in the wild. We actually cherish the birds both in captivity and in their wild state. Many aviculturists contribute to conservation programs for the preservation of a species. I feel that government conservation agencies have improved their concepts in recent years in Australia, although they are still very reserved in their attitude towards aviculture and ultimately this is only of greatest detriment to the birds themselves.

Birds have been confiscated from trappers and released back into the wild. The trappers were obviously in the market to make money and deserve to be prosecuted. However, it is abhorrent to me to hear of Major Mitchell's Cockatoos, which were taken from the nest and handreared by the trapper, being released back into the wild on confiscation. These birds have not been educated by their kind to be able to survive in the wild and therefore face a very cruel death through starvation or hopefully a quicker

Red and Blue Lory.

death from a predator. It sickens me that a conservation agency can employ apparently well-educated people to be so callous, consciously or otherwise.

I believe that aviculture can make an extremely important contribution to conservation and in the future this will become even more apparent. In Australia, a country noted for its capabilities in breeding species, we need to be able to support conservation.

Third World countries cannot afford to keep birds and in many cases they need to eat them or sell them to survive. We cannot tell a person in Indonesia that it is wrong to catch a particular bird, if it is his sole means of putting food on the table. I believe that certain species should be trapped for avicultural conservation when a particular region is logged. We will not stop deforestation unless we prevent the demand for the timber and prevent the need for agricultural land. I am sure that if local people were educated to understand that the land in its original pristine condition can supply them wealth, they will not disturb it, but utilise it. Controlled wildlife harvesting is required and is becoming frequently an acceptable reason to conserve biodiversity for the future. Papua New Guinea has programs in place where rare Birdwing Butterflies are being utilised to provide income to extremely isolated communities. These people are educated to plant and protect a species of vine that the caterpillar needs to eat. Birdwing Butterflies are actually regaining some of their original habitat because a Third World country is seeing the light. These butterflies are collected and sold to the government for sale to butterfly collectors worldwide. Not all butterflies are harvested and although the project is in its infancy, butterflies are again flying through a region from which they had disappeared.

Australia could probably establish viable conservation projects through aviculturists to protect birds from being swallowed up in land clearance. Enough has been said in regard to the struggle for conservation. I only hope that we can all work together for the wonderful and diverse species of fauna and flora on our planet, as the future is bleak for them.

Very few records are available to determine how common or uncommon species of lories and lorikeets are in their native habitat. *The Atlas of Australian Birds*, a Royal Australian Ornithologists Union (now Birds Australia) publication, offers an indication of density of Rainbow Lorikeets in suburban Townsville, North Queensland. The density is 4.1

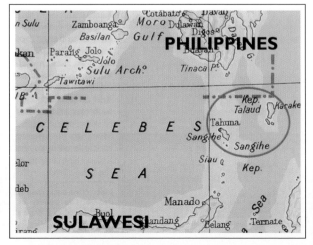

birds per hectare and the greatest numbers occur between Cape York and the mid-north coast of New South Wales. Most large flocks are seen within 100km of the coast. Using this area and the given density we could calculate that this region contains approximately 93 million birds. Obviously, Townsville does support larger numbers than many areas within this vast region and no doubt this figure can be reduced to a more realistic factor, but even a 50% or 70% reduction still indicates that Rainbow Lorikeets are numerous.

We must not, however, estimate all species to be this abundant, as some are certainly vulnerable or endangered. The Ultramarine Lory *Vini ultramarina* of the Marquesas Islands is certainly worthy of concern. It is estimated that pairs number in the hundreds and not thousands, on the islands within the group. Recent work by the staff of the San Diego Zoo and a delegation from the Department of the Environment for French Polynesia has involved moving some captured birds to the island of Fatu Hiva from Ua Huka.

Ultramarine Lory.

Habitat destruction or the introduction of a pest such as the black rat will see the birds exterminated. I hope that some pairs will be taken into captivity to be bred by knowledgeable lori-culturists and thereby help support wild conservation of the species. Zoological input has so far helped with the conservation of the Californian Condor in the USA and I am sure that in time the Ultramarine Lory will establish in aviculture. It has already proved to be capable of reproduction in captivity by the Duke of Bedford. Aviculture has progressed positively since that breeding in 1939, and with our support in such a program, Australian aviculturists could do for the Ultramarine Lory what has already been achieved with the Cuban Finch. If not for Australian aviculture this species of finch would likely be extinct. They are indeed so common these days that you can buy them for approximately $AUD30–40 a pair.

The Red and Blue Lory *Eos histrio histrio* of Sangihe, Talaud and Nenusa Islands between Sulawesi and Mindanao in the Philippines has recently been heavily traded. Birds on Sangihe have been all but exterminated (possibly extinct) due to habitat loss for coconut and other agricultural plantations. However, birds available in the trade now are mainly from the Talaud groups of islands. I have seen hundreds in Singapore, but to date breeding achievements are occurring mainly in Europe, and I can only say that the large numbers captured are not justified and cannot be supported from the estimated remaining wild population. Future research on the status of the Red and Blue Lory may or may not condemn the large numbers being trapped for the bird trade. Since the first edition of this book programs have been developed to educate the local population. This education program is similar to those developed for the Caribbean Amazon Parrots. Hopefully the locals in time, will see and understand the benefits of leaving the birds alone.

I do not support uncontrolled trapping of a species, however, if habitat destruction is occurring, controlled capture before logging must benefit the species. Once an area is logged, the animals from that area infiltrate adjacent untouched areas causing extreme pressure on existing flora and fauna. Such biological upheaval is unnecessary and could be avoided with a better comprehension of the basic laws of nature.

Some interesting records obtained through the Traffic Bulletin, and set out in the table, indicate the extent of exportation of species of lory from Irian Jaya, Indonesia.

RECORDS OF IRIAN JAYA PARROTS BETWEEN 1985-1992

SPECIES	CAPTURE PERMIT	SHIPPED	CITES DATA
Black Lory	3120	1551	2318
Duyvenbode's Lory	6280	2181	3216
Yellow-streaked Lory	1169	11428•	1359
Josephine's Lory	6018	1227	1153
Striated Lorikeet	75	–	395
Papuan Lory	6570	3023	4306
Red-flanked Lorikeet	3069	136516•	2797
Fairy Lorikeet	4759	1901	2613
Red-fronted Lorikeet	–	–	499
Wilhelmina's Lorikeet	175	–	134
Black-winged Lory	2766	1074	968
Violet-necked Lory	225	13	4215
Black-capped Lory	–	25	2
Musschenbroek's Lorikeet	7115	1827	2051
Emerald Lorikeet	–	135500•	234
Whiskered Lorikeet	850	326	1390
Dusky Lory	9479	3768	5444
Goldie's Lorikeet	5499	2752	2167
Rainbow Lorikeet	12920	5933	32659

* *These figures indicated in the table are extremely inflated by comparison to any other species. Emerald Lorikeets are very scarce in captivity and a figure of 135,500 for shipping is ridiculous when compared to the figure of 1827 for the Musschenbroek's Lorikeet which is much more plentiful in captivity.*

    Unfortunately, these figures can be misleading in many ways. Questions which arise are obvious. Have there been enormous losses between capture and shipping, or have some been unrecorded? If these values are accurate, can certain species sustain these figures, eg island species such as the Black-winged Lory *Eos cyanogenia*? A figure of 32,659 *Trichoglossus* species over this period of time is, I feel, easily sustainable. The other question that arises is how many of these were Rosenberg's Lorikeets *T. h. rosenbergii*, a subspecies of the *Trichoglossus* genus found in a restricted habitat and possibly vulnerable to trapping?

    The answers to all our questions lie in further research of the status of a species or subspecies and stricter control of permits which I understand is extremely difficult in Third World nations. Let us face it, Australia cannot control illegal trapping effectively even with legislation in place. This legislation requires permits to keep most Australian species. Many are still traded illegally within Australia and of major concern are species smuggled out of the country which are rare, such as the Golden-shouldered Parrakeet. Captive bred eggs being smuggled are, although still illegal, nowhere near as detrimental as smuggling live birds in terrible conditions, particularly if they are wild caught.

    It is most unfortunate that some of these birds are wild caught and we could certainly prevent the strain placed on wild populations by controlled export of aviary bred birds. Conservation is the concern of every individual who lives on our planet. We all share the earth with fauna and flora which have as much right as humans to exist. It is important that aviculture continues to support conservation and it is important that we do not threaten wild populations. We can coexist with wildlife, and we must, for as many future generations as possible.

LORIES AND LORIKEETS IN CAPTIVITY

Pair of Scaly-breasted Lorikeets

LORIES AND LORIKEETS IN CAPTIVITY

People have no doubt tried to keep these colourful birds in captivity since the early explorers first discovered the Spice Islands. Unfortunately, due to their feeding requirements, most birds would have succumbed in a very short time. It is only in recent times that our knowledge has afforded us the ability to keep and even breed most species.

It must be remembered that there is no point in deciding to keep a bird in captivity if you are not prepared to give it adequate accommodation and the correct nutrition for its well-being. It often amazes me to visit a bird keeper who in many cases has spent a considerable amount of money to obtain a particular species. However, the unfortunate birds are often accommodated badly because the bird keeper is not prepared to offer adequate accommodation on the grounds that it is too expensive. Please note I use the word bird keeper and not aviculturist.

Once you have decided to keep a particular species it is vitally important to try to breed the species. We have no right to keep birds without giving them the opportunity to reproduce. Aviculture should now be considered a form of preservation and not just a hobby. It is essential for species to propagate in captivity to ensure the survival of a species and to contribute to future conservation needs. Gone are the days of losing a bird due to ill health and replacing it by capturing a wild specimen.

Aviculture needs to self perpetuate a species and should not rely on wild stock in the future. We, in turn, can contribute to the conservation of a species by making aviary bred stock available for re-introduction into the wild. Re-introduction is no easy program to accomplish but we need to be able to try to sustain wild populations where habitat is still available to justify release.

Selection

Suitable birds can be obtained from breeders and commercial establishments. It would obviously be a great advantage to your purchase if you can establish a history of the birds before you acquire them. You may be lucky enough to discover this history by visiting the breeder. Many aviculturists try to avoid buying birds from a dealer for various reasons. However, there are many reputable dealers from whom you can confidently purchase birds, just as there are many breeders whom you cannot trust in regard to obtaining a legitimate history.

In the end you will have to make the final decision. Following are some suggestions to help you. Research the species before you visit a dealer or breeder premises. Ask the age of the

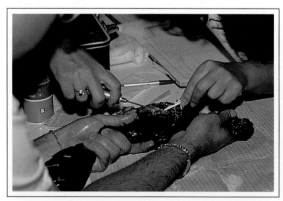

Above: Microchip implanting at Loro Parque.
Right: Surgical sexing at Loro Parque.

birds. Are they an unrelated pair? Have they bred? What are they being fed? What size accommodation have they been kept in? Are they local or have they recently come from a different climate? Many breeders and dealers offer surgically sexed pairs. Ask for their veterinary certificates. Look at the birds and check if they have clear and bright eyes. Is their plumage in good condition? Do they have all their toes? Do they have a clean vent? Do they have nasal discharge? Are they breathing naturally? There are many questions you can ask and answer yourself, but once you are experienced these do not need to be memorised, they will just come naturally when deciding on a purchase.

Sometimes your selection may involve selecting a pair that are not surgically sexed and your selection may be from a group. Generally speaking most parrot cocks are slightly larger in build and more forthright in their behaviour within a group. Please note however, that there are always exceptions to the rule. In general terms I would use the following principles in selecting a pair. Watch the group and look for pairs within the group. Often pairs are already formed. If the pair are mutually preening and follow each other around when disturbed, the chances are that they might prove to be a compatible and eventual breeding pair. Be aware, however that cocks will often associate with other cocks and hens with other hens, so this is not foolproof.

If I felt that the compatible birds visually looked like a pair, I would somehow mark them and then request their being caught. A squirt from a water spray with some dye or some physical characteristic such as a bent tail may help mark your pair.

Once caught, hold a bird in each hand if possible and compare the head size, width of cranium, and size and width of mandibles. I have found that cocks are generally larger in these areas and also tend to be slightly brighter in colour than hens, even in lory species. Remember if you have no definite knowledge of age then this is certainly not conclusive. An older bird tends to be of a brighter plumage than a younger bird. While handling the birds observe their health and fitness by feeling the breastbone. A thin bird may have health problems that may be impossible to remedy. A bird which is fluffed up or has its head placed over the shoulder onto the back and particularly has two feet on the perch, should be treated with suspicion. It is likely that the bird is not well due to disease or stress. If no pair bonding is apparent within a group, it is best to select the largest bird of the group and pair that with the smallest bird. Again comparing the head size, cranium width, mandible size and width, will usually prove you have selected a pair. Obviously, surgical sexing though not foolproof, particularly in young birds, is the preferred way of defining a pair. DNA blood and feather testing is also widely used for sex determination.

A bird with dirty feathering does not necessarily mean an ill bird. Lories and lorikeets are messy eaters and when kept confined with others, particularly in a dealer's premises where bathing opportunities are minimised or impossible, a bird may become grubby. If a grubby bird is selected, make sure that you check for bright eyes, clean vent, no nasal discharge, natural breathing and good muscle development around the breastbone.

Transportation

You are now the proud owner of two new birds and you need to transport them to their new accommodation. Make sure they have a clean sanitised box large enough for them to travel comfortably. It is often wise to separate the pair into different boxes or compartments as they are under high levels of stress and may inflict injury upon one another if upset.

Some birds may need a padded box although this is extremely unusual. If the birds are flighty while travelling, cover the box with a towel. If you cannot be in close contact make sure that the transport box is enclosed to facilitate privacy so that the birds will not notice much of what is going on about them.

Make sure that draughts do not affect them, but at the same time allow for adequate ventilation. Do not leave birds in the boot or car in hot sunlight for any length of time.

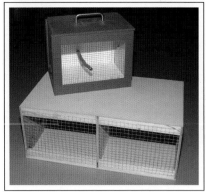

Carry and transportation boxes.

Heat exhaustion will soon kill any bird.

Depending on the duration of travel you may need to feed the birds. On short trips of only a few hours there is no need to feed. On longer trips of up to 24 hours provide a dry mix (discussed in the *Feeding* section) and fruit, preferably an apple. The fruit will supply enough moisture in a 24 hour period. Wet mix usually contaminates the bird's feathering and is unnecessary.

On longer trips it may be necessary to stop and feed the bird with wet mix and water while you are having a break. Once fed, I would then remove the dishes before continuing the journey.

It can be useful to add a vitamin or electrolyte supplement to the drinking water to combat stress.

Do not introduce the birds to their new surroundings late in the day. During the night if they are unfamiliar with their new accommodation, they may fly and injure themselves, often due to the least possible disturbance. Leave them in their transport box with food until the next morning. This will give the birds a whole day to explore and familiarise themselves with their new surroundings. They will then be prepared for the first night in their new accommodation.

Clean your transport boxes and keep them in good condition. It is advisable to keep new birds quarantined from other birds until you are satisfied that there will be no risk of disease transferred into your existing collection.

Feeding

I could write a book on the various recipes prepared for and given to lories and lorikeets, but this will only serve to confuse a beginner to lori-culture.

I do not claim to be a nutrition expert, but I can offer my feeding program with the knowledge that I have successfully kept and bred all Australian species and many exotic species.

There are many commercial products on the market and manufacturers all vouch for their nutritional content. I am in no way refuting these products, as I have tried many of those available in Australia and overseas. I personally used to supplement my birds' diet with Lorinectar™ and Loristart™, both imported from Holland. These days I feed these two products as the maintenance diet. I find them convenient to mix and proven products after using them for six years.

Choose a reliable, respected and proven brand if you wish to purchase a ready-made formula.

I feel that my previous diet was adequate, but I am sure my birds appreciate a change every now and then as we do.

I believe it is important to feed both wet and dry mixes although lories and lorikeets prefer wet mixes. When lories and lorikeets eat a dry mix

The author feeding a pair of Black Lories and their young.

they fly to water to 'make' a wet mix. There is usually a lot of dry mix in the bottom of the water bowl. Feeding birds on a dry mix alone proves costly, as the birds do not access this layer at the bottom of the water bowl. Often they will also scatter the dry mix onto the floor while feeding or out of wanton delight. In my opinion it does not solidify their faeces due to their immediate need for water when feeding on dry mix. Wild lories and lorikeets may feed on pollen and at certain times of the year a larger proportion of 'dry' pollen will be consumed. They still squirt their faeces as this is their natural way of passing waste.

Above: Fresh fruit & vegetables are supplied daily. Below: Purple-crowned Lorikeet at stainless steel feeding bowl.

My previous wet mix was prepared in the following way:
1 cup of Farex™ baby cereal
1 cup of finely blended plain biscuit (eg Arnott's Nice™ biscuits)
$1/2$ cup of wheatgerm cereal
1 tablespoon malt powder (Nestles™ brand)
A pinch of table salt

These ingredients are thoroughly mixed and placed in an airtight container.

Each morning I made a fresh mixture as follows:
$1^{1}/2$ cups of the above preparation
4 dessertspoons of blended vegetable mix (can of Edgells™ vegetable mix which contains potato, corn, carrot and peas)
2 dessertspoons of honey alternated with 4 dessertspoons of brown sugar on alternate days.

I add approximately 400ml of boiled water to help disperse the honey and then add cold water to make up 1.5 litres. The Farex™ and biscuit combination is added and stirred. The 1.5 litres serves 14 lories. Chattering and Red Lories receive 250ml per day per pair. Medium-sized birds, ie *Trichoglossus*, receive 200ml per day per pair. Small-sized birds, ie *Glossopsitta*, receive 150ml per day per pair. As I have previously stated I had much success with this formula.

During certain periods, such as when young are in the nest, the amount may be increased accordingly. At certain times of the year it may also prove necessary to decrease (to prevent unnecessary waste) or increase the quantity. For example, in winter, pairs will consume more. This mixture is warmed and the birds eagerly await and partially consume the mix as soon as it is placed in the aviary. During the day it cools off but the birds will finish the mixture.

Lories and lorikeets will sample all kinds of food and will eat vegetables such as celery with relish. Fresh sweet corn on the cob is not consumed unless the birds have learnt how to break through the skin to access the milky heart. If frozen sweet corn is purchased and defrosted the corn is juicy and the birds look forward to this treat. It is particularly useful when young are

Loristart™ and Lorinectar™ are complete formulas and easy to prepare.

in the nest. A pair of Musk Lorikeets I owned, seemed to be only content when they received their corn each evening. Fruit such as sweet apples, pear, mango, pawpaw (papaya), grapes and melons are readily consumed. In my experience, bananas, apricots, nectarines, plums and peaches are not as acceptable. Each evening when young are in the nest, I feed sponge cake that has been lightly moistened with sugar water. This is also given occasionally, say once every two to three weeks, out of breeding season.

Calcium in the form of cuttlefish is usually unattractive to lories and lorikeets, therefore I supplement the diet by adding Calcium Sandoz™ syrup (22mg calcium/ml solution), MSA (Nekton™ products, Germany) or any other calcium powder to their wet mix. This is done more frequently prior to breeding to assist in the formation of the eggshell, but is also given occasionally during the non-breeding season.

Vitamin E powder (Nekton™ products, Germany) or wheat germ oil can be added to the wet mix to increase fertility. Vitamin drops such as Abidec™ or Pentavite™ are rarely added to the wet mix as I feel this formula already caters for most vitamins adequately. When using Lorinectar™ and Loristart™ which are complete formulas, I do not add calcium, minerals or vitamins.

Fruit or vegetables are given each evening. Try not to waste feed. The birds should get just the right amount to meet their needs. Dry mix as described by Stan Sindel is given to the birds and is available at all times. I feed dry mix for two reasons. Firstly, it gives them a choice and secondly, if I ever happen to be unavailable to feed wet mix, the birds will certainly be adequately maintained.

The mix as detailed by Stan Sindel is:
2 cups of rice cereal (Heinz™ or similar)
2 cups of rice flour
2 cups of Canary egg and biscuit food
1 cup of glucose powder
1 teaspoon of multi-vitamin and mineral powder (Sustagen Sport™)
1 dessertspoon of pollen (optional)

I have never added pollen because it is rather difficult to obtain regularly and is expensive. Flowering non-poisonous sprays can be given as a treat and eucalyptus, grevillea and bottlebrush (*Callistemon*) are particularly enjoyed. The best time to feed this is when nectar is seen on the flowers. Fresh leaves and thin stems of shrubs and trees are also vigorously chewed and played with. Many lories and lorikeets will chew the leaves or bark and then run their beaks over their feathers in a grooming motion. This occupation, known as 'anting', can be immediately observed and certainly promotes immediate interest when fresh branches are placed in the aviary. No doubt this benefits the birds although we do not exactly know why. It may be that oils obtained from the leaves or bark may repel mites, but this is only an assumption.

I have not gone into the requirements of amino acids, vitamins and minerals in this section as I believe you could write a book on nutrition alone. The mixtures I feed have produced healthy self-perpetuating stock and I feel that nutrition should not be tackled by a novice. The baby cereals we use are formulated for humans and contain the correct ingredients when supplemented with fruit and vegetables to produce and maintain healthy birds. Lorinectar™ and Loristart™ however, are better formulated for feeding lories and lorikeets.

Housing

Although seldom considered, it is important to analyse your requirements and more importantly that of the birds. There is no room in aviculture for the permanent housing of a species that does not allow the bird to move freely, communicate and breed.

A happy active bird is much more likely to breed than a bird that is influenced by a stressful environment. Select the species you would like to work with, then learn its requirements. The next question to be asked is whether you can afford to house this species and create a suitable environment. Anyone can keep a bird in captivity, but please remember that you have a responsibility to that bird, and if you do the right thing you will reap the benefits. No species should be kept in accommodation that does not allow it to fly, clamber or observe what is going on in its surroundings.

I do not believe in permanent cages. Birds can be housed satisfactorily in suspended aviaries (not cages) and conventional aviaries.

Only one pair of lories and lorikeets should be kept in each aviary due to their aggressive nature. Small species such as the Musk Lorikeet should be kept in a suspended aviary with the minimum dimensions of 2 metres long x 90cm wide x 90cm high. Larger species such as the *Trichoglossus* genus should be housed in an aviary with the minimum size dimensions of 3 metres long x 1.2 metres wide x 1.2 metres high.

When species are housed in larger

Above: Inside aviaries – Holland.
Left: Feeding door on suspended aviary.

aviaries, it is amazing just how swiftly some species do fly. It may even be necessary to place some form of noticeable barrier, such as sacking or fresh cut leaves, at the end of the aviary when young leave the nest. Young can react in an irrational manner when recently fledged and do take fright easily until more accustomed to their new surroundings. The same can be said for newly purchased birds unfamiliar to your aviary.

I find that 25mm square galvanised steel tube is the best material for the construction of aviaries. Tube can be cheaply welded and in the case of furniture tubing, special fittings are available. These fittings are obtainable in all the shapes required to build your

Above: Suspended porch aviaries – Belgium.
Below: Conventional aviaries – Australia.

aviaries. They come in plastic or metal and are expensive, but easy to assemble. Make sure that the plastic fittings are resistant to ultraviolet radiation if they are to be used in direct sunlight. If they do not have an ultraviolet inhibitor, the plastic will become brittle over time and your structure will eventually break at these joints. If you intend to use this principal form of construction, it is more expensive to use the metal fittings but eventually it will be more reliable. Furniture tubing is a thin gauge material that may not be long lasting or structurally sound under certain conditions, so please be aware of these shortcomings.

Beware of galvanised wire that has been poorly manufactured. Zinc flakes and other residue, once eaten, can kill your birds due to zinc poisoning. Powder residue is generally not as serious, but can cause illness and eventual death. Some species, such as Hooded Parrots, will suffer losses as they are prone to chewing on wire. However lories are not as susceptible. Do not take the risk. Clean your wire with a file, a wire brush and vinegar to remove all excess particles and as much powder residue as possible. Even better still, refuse wire that is not up to a satisfactory standard. If more customers refuse poor products, then manufacturers might improve their quality control and hopefully provide wire that does not need this labour intensive work before use. One Steel's (BHP) product, Evencoat™, has been developed as a direct result of the problems associated with 'new wire disease'. The post galvanising process all but eliminates the lumpy residue that was common with wire mesh. Vinegar wire wash is still recommended followed by painting the wire with an oil-based dark coloured paint which allows for a far better viewing of birds.

Lories and lorikeets do not require heavy gauge wire, 1.6 gauge is adequate and 12.5mm square welded mesh is satisfactory. This size will keep out larger pests, although it is amazing how a mouse can actually still squeeze through this small aperture. Complete control of mice can be achieved using 6.25mm square wire, however the birds are difficult to observe through this material and it is expensive.

In Australia the use of wire larger than 12.5mm square will allow a variety of pests and predators to gain entry to the aviary. Snakes and rodents are unwelcome visitors which must be prevented entry. The use of flat Colorbond™ steel sheeting around the bottom of conventional aviaries to a minimum height of 60cm will prevent snakes and rodents climbing up and through the wire. It is best to turn the top of the sheet away from the aviary and down at an angle towards the ground. If a pest tries to climb the flat surface it will usually be unable to overcome this obstruction. Electrical wire can be run along this surface to further protect your birds. A simple system using an agricultural electrical fence as used for cattle and sheep can also be purchased. Please remember that although overhanging shrubs and trees look great surrounding your aviary, they are also a stepping stone for pests. There is no point in having your aviary protected from below if a snake can drop onto the aviary from an overhanging branch above.

Concrete foundations should be placed under the aviary to a depth of at least 20cm, deeper in sandy soils and shallower if you have hard rock conditions near the surface.

Colorbond™ sheeting can be used to protect the sheltered portion of the aviary. However, please remember that in summer these shelters can become extremely hot and in winter condensation can occur. Birds seen panting in hot weather are usually in a stressed state which is entirely unnecessary. Insulating the roof is essential, especially in hot climates like those experienced in Australia. I recommend a manufactured sheet with a roof profile on one side, flat Colorbond™ sheeting on the other and an inner insulation of polystyrene foam measuring 50–75mm in thickness. This specially produced insulating material is used in the manufacture of coolrooms and the like. It is expensive, but reject or damaged sheets can be purchased and will be appreciated by your birds. At least a third to a half of the aviary should be sheltered. Lories and lorikeets usually prefer to roost in a nestbox during the night and of course this gives added protection and comfort to the birds.

Natural perches should be placed at either end of the aviary. One should be situated under the shelter and placed higher than the flight perch, to entice the birds to roost in the shelter. Birds usually prefer to sleep on the highest perch in an aviary.

I recommend natural perches. Make sure that they are not too big or too small in diameter. This allows the bird to sit comfortably with its toes gripping the perch. It is also wise to vary the thickness of the perches so that birds can exercise their toes on twigs and larger branches. They will usually select the most comfortable section of the perch to roost on during the night.

Fresh perches should be renewed regularly and it is best to devise a simple system of removal and replacement. Birds chew the bark off fresh branches, so make sure that you do not place poisonous plants in your aviary.

A large, shallow container of water should be placed away

Conventional aviaries at Jurong Bird Park – Singapore.

Suspended aviaries at Walsrode – Germany.

from perches to prevent fouling. In small aviaries fouling may be difficult to prevent as lories are able to squirt their faeces. Supply fresh, clean water as frequently as required as lories and lorikeets love to bathe daily. The depth of the water should be such that it covers the lower abdomen of the bird and can be fluffed over the upper body by the bird's wings. If the water is too deep the bird will be afraid to enter and will not fulfil its bathing requirements.

Stainless steel food bowls are my preference as they are durable and easy to clean. They should be accessible from the outside of the aviary from an escape-proof alleyway (safety porch). The need to enter a lories' territory to feed, can disturb the birds and may influence breeding results. It is also easier to access an aviary from the outside. Some species become rather cheeky and will often bite unwary fingers so access via a swivel door will prevent painful bites.

To prevent contamination all bowls should be cleaned every day before the daily wet mix is added. Dry mix bowls can be left longer, but must be regularly inspected for fungal growth on the sides and should be cleaned when necessary.

I suggest collecting bowls each evening and placing them in a trough of hot water with detergent and a chlorine (bleach) solution. Leave bowls overnight to soak and thoroughly brush and rinse in clean water every morning.

The aviary adjoining food areas may become spoiled with food particles that lories often flick off their bills. It will be necessary, once a week or in cooler weather once every fortnight, to clean this area. Spray some water from a garden hose all over the aviary and fouled food areas. This allows the soaking of hard, dried crusty areas. Finally, clean the aviary with a high pressure spray.

Running drippers into bathing facilities and spray systems on top of aviaries is up to the individual aviculturist, however this will be greatly appreciated by your birds.

It may also be worthwhile to fully cover the wire roof of your aviaries to prevent

The author's aviaries.

the introduction of disease from birds perching on top. Droppings from infected wild birds could prove a catastrophe for your birds. It is a pity that birds will not be able to 'rain bathe' under these conditions, but they can be lightly sprayed regularly using a garden hose or watering system. The use of shadecloth may be a valuable addition to keep species cool in summer and help prevent faeces from wild birds entering aviaries. Doors should be sized to allow easy access for the aviculturist and prevent birds escaping. Safety porches at the front of or behind aviary complexes are essential. If birds do escape, they can be ushered back into their aviary. Design a system that prevents unnecessary stress to the birds should this happen. Some aviculturists occasionally use these walkways to provide their birds with additional exercise. If this is considered, remember that a pair could be attacked or will attack other pairs through the wire and may be severely injured.

If the sides of aviaries are used as barriers between adjoining pairs remember to make them double wired with at least 25mm separating each wire partition. You may think that adjoining pairs have accustomed themselves to their neighbours' aggression, and that less aggressive birds will avoid contact. This may justify single wire partitions and be economical in your mind. However, once young leave the nest they invariably hang on the wire with their beaks and feet. At this point they are unaware of any impending danger and incredible damage can be done by the adjoining pair. Do not risk it.

Security is another aspect to be considered when building aviaries and I certainly recommend that you consult the experts. Unfortunately, more and more thefts are occurring within our civilised and overpopulated communities. I have opted for the use of pressure pads and infra-red switching devices. These can, when triggered, light up your aviaries and trigger alarms to ring, scaring the intruder or alerting you inside your house.

Another form of security I employ is the microchipping of birds. If a theft does occur and you are successful in locating the stolen birds, you will need to prove your ownership. A simple wave of the scanner will convince the authorities that you are indeed the owner and the thieves or the unfortunate new owners will be required to return the birds.

It is amazing that when rare birds are stolen they will eventually surface and can be detected through the aviculturist and dealer channels. Inexpensive birds may be much more difficult to locate.

Breeding

Prior to the breeding season increase the calcium levels to assist in egg production. Using Calcium Sandoz™ syrup (22mg calcium/ml solution), add 10ml per litre of wet mix once per week. When using other calcium products add their recommended dose rate to the wet mix. I also recommend adding Vitamin E to encourage high fertility, observing recommended dose rates. As mentioned previously under the section on *Feeding*, Loristart™ and Lorinectar™ have sufficient calcium for my birds' needs.

Nestboxes should be clean and bedding renewed. Use wood shavings without fine particles. Fine

Suspended aviary housing Whiskered Lorikeets with young at Walsrode. Note nesting log.

wood dust can cause death in young chicks. Be sure to use untreated shavings as much of our timber is treated chemically and may be harmful to your birds. Replenish the wood shavings when the nest is fouled. My birds are used to having the nestbox removed and the young handled. The nestbox may even be cleaned with a hose and the young placed in an interim box while the nestbox is dried. The young are then put into the dry box with clean shavings added and then reintroduced to their parents. With tolerant parents this procedure is acceptable.

If your parents particularly resent intrusion, it may be necessary to lift the youngsters out and just place clean shavings on top of the existing fouled bed litter, quickly replacing the chicks. It is usually best to perform this function when the parents have voluntarily left the nest. Many elaborate nestboxes can be developed to keep the nest litter clean. If young are left to stand in the ooze of droppings they can become cold or diseased and will succumb.

Due to the liquid nature of droppings I recommend nestboxes constructed from marine grade plywood with a minimum thickness of 12.5mm. On the inside of the box fix a weldmesh ladder to allow the parents to clamber in and out without jumping on and damaging eggs. All sharp wire cuts should be filed smooth so that no damage can occur to the chicks or parent birds.

It may be necessary to insulate a nestbox in cold climates using a layer of polystyrene, creating a double box system. Another method to keep chicks warm when the parents no longer brood is to place a 40-watt globe at the side of the box. This is sufficient to keep young warm as they will move towards or away from the globe as desired. If a globe is placed underneath the nestbox chamber, the chicks have no way of escaping the heat. If a hot day occurs while you are absent you may find all the young dead on your return.

In the subtropical climate of south-east Queensland, this globe system is generally not necessary, however Varied Lorikeets would be the exception as they breed in our winter. Parents do not seem to bother with

Top: *Variety of nestboxes suitable for lories and lorikeets.*
Above: *This weldmesh ladder minimises the chance of parents breaking eggs when entering the nestbox.*

Fostering Cardinal Lory chick to Dusky Lories.

brooding after approximately ten days. At this time the young have virtually no down and will succumb very quickly in cool winters. The parents may even enter the box at night but may not brood the young while roosting. Take care to understand what all your pairs are capable of doing, as they can differ in their requirements.

Usually young birds are independent of their parents two weeks after leaving the nest. I have observed some young already trying to sample wet mix and fruit with their parents on their first day of leaving the nest. Fledglings learn very quickly, but I prefer to wait until at least two weeks and be sure of their independence. Young can be left with parents for longer periods as most parents are tolerant of their young. However, should you observe any aggression from the parents toward their young it would be advisable to remove the young immediately.

When young are left for long periods with their parents, they can influence any further breeding. I have had pairs lay and incubate eggs with older chicks still sleeping in the nestbox. Eggs have been damaged, but I cannot say whether this was on purpose or by accident. Young that have hatched have also been found bitten. I cannot be sure if the older chicks were responsible or the parents may have done so under stress, because the older chicks were still in the nest.

If you want to continue breeding remove the young. Colony breeding has been attempted successfully by some breeders. In my experience there has been interference with dominant pairs breeding and lower ranked pairs being much less productive, or possibly not breeding at all. In fact, successful colony breeders may find that if their pairs were housed individually their reproduction would also increase. This applies to smaller species as well as the larger forms.

It seems that breeding results for some of our exotic species are inconsistent and in some species we have considerable difficulties. The Yellow-bibbed Lory comes to mind. Speaking to many aviculturists who keep this species, they have found that their pairs destroy eggs, crack eggs and do not raise their chicks. Is this due to all the handrearing that is being carried out on rarer species? Are our pairs losing the capability of natural breeding due to our interference with their parents or their previous clutches?

I think it is vital to breed naturally with your lories or at least give them the opportunity to rear some clutches.

In speaking to European and South African aviculturists who are breeding from imported wild Yellow-

Yellow-bibbed Lory chick.

Handfeeding Purple-crowned Lorikeet chick.

bibbed Lories, they are finding them reasonable performers. All Australian birds are from a captive background and indeed many are handreared.

Those Australian aviculturists with parent reared pairs or natural breeders should keep up the good work. Their young should be advertised as such, as I believe they are worth more than handreared young.

Handrearing

More frequently these days, with our increased knowledge of the requirements of captive birds, we are capable of assisting reproduction. In the wild many tragedies occur within the nest as has also been the case in captivity. A pair that lay fertile eggs but do not incubate or may not feed their young adequately, can still be used to produce healthy offspring.

Young, handreared birds, which have never seen their parents, can be capable of laying fertile eggs and rearing their own young without assistance. I do believe that we may take this a bit too far, and to increase production and sales may eventually influence our birds to require human assistance. It has happened with certain quail and other domestic species and these species are becoming more difficult to breed naturally. It is my firm belief that a pair of lories should be given the chance to raise their own young. Certainly increase production, but limit your handrearing to allow the birds to parent rear at least one clutch per season.

Living near the coast in subtropical Queensland, I incubate my eggs at 37°C and place a cup of water in the incubator for humidity. The humidity stays between 45–55%, which is ideal. It may be necessary to be much more fastidious about humidity in drier or wetter climates, but please remember that I am relating my personal experiences. These may need to be modified to suit your conditions.

The eggs are turned approximately every two hours. Each egg is carefully rolled through 180 degrees to its new position for two hours and then returned to its original position two hours later. Do not worry about turning the eggs during the night. I have had great success by turning the eggs on average 7–8 times per day. Do not become paranoid and start assisting the chick in hatching. Even those with experience have unwittingly killed a chick, because they were worried and wanted to help it. The reality is that the hatching chick is robust and generally capable of doing very well without your interference. Once the internal pip (determined by candling) has occurred, do not turn the egg. Within 48 hours the chick should have hatched. It is my opinion that you should not worry until 48 hours have elapsed. It may then be necessary to assist the chick to hatch. If you are inexperienced in this procedure travel with caution, read all the books or better still ask an experienced aviculturist or avian veterinarian to help you. Once the chick has hatched do not sever the umbilical cord. Allow it to dry naturally and it will soon fall off.

The chick does not need immediate feeding and some breeders wait for 24 hours. I have successfully raised chicks that I have commenced feeding after a four hour interval after hatching. No doubt others have fed even earlier but there certainly is no need. During the first 24 hours of feeding, the chick is fed a dilute soupy mixture of two parts water to one part beef and vegetable broth (as used for human babies eg Heinz™ or

Gerber™ brands). I use a spoon with both sides bent up. It may be necessary to feed with a pipette or syringe, however I generally have no trouble using a spoon from the first feed. The temperature of the mixture should be tested on your lips. It must not be too cold and if your lips can take the heat, it will be acceptable to the bird without discomfort or scalding of the crop.

Scalding of the crop is inexcusable and you do not need a thermometer to prevent it, just your lips. I have had breeders tell me that the young will only eat the mixture at 50°C which is far too hot. If they stopped to think, they would realise that a bird with a body temperature of approximately 38°C cannot bring regurgitated food to a temperature of 50°C. Birds do not have an electrical device in their metabolism.

The chick is kept at a temperature of 37°C for the first one to two days and its two hourly feed intake is merely a couple of drops off the spoon. After 24 hours I add a small proportion of my wet mix (the dry ingredients) to the water and beef and vegetable broth mixture that is still very watery. The water used in all my recipes is filtered to exclude most of the chemicals, which have been added to our suburban water for purification. This mixture is gradually thickened to provide more nutrition and less water in the mix. By day four the chick will be eating a mixture with a porridge-like consistency and is fed this until weaned.

From about day three the mix is:
1 part wet mix (ie Farex™, blended biscuit, wheat germ stored dry)
1/2 part blended apple (tinned puree)
1/2 part beef and vegetable broth

There is no need to introduce micro-organisms to the crop as often suggested. I have done this in the past, but do not usually bother as the youngsters progress very well without it. Do not overfeed, and let the crop empty before the chick's next feed. This will indicate whether the contents are being absorbed as crop stasis (sour crop) is a problem that must be controlled. The occasional drop of Pentavite™ or Abidec™ on the spoon is given say, once a week. After about a week I occasionally add a small amount of sponge cake to the mixture.

Food is heated in a microwave oven and stirred thoroughly to produce an even temperature throughout the mix. Microwave ovens have hot spots if you do not stir, so be careful that you do not scald your chick or more correctly your lips.

A mixture, sufficient for approximately one day's use is prepared, and after each feed is placed into the refrigerator. In recent years I have used Loristart™ very successfully. This is not to say that what I used previously is not successful. However, due to increased knowledge and through experimentation I now feel that this product provides everything that the birds require. Growth rates are excellent and I have found that crop stasis is very infrequent. With my old recipe, I frequently had some crop problems.

I use apple cider vinegar to clean crop stasis. Add 1–2 drops to a dessertspoon of water and bring to feeding temperature (between 37.7–43.3°C). Generally only one administration of

Handfeeding Dusky Lorikeet chick.

this solution is required to ease crop stasis. Any food in the crop is then consumed and you can then resume your usual feeding regime. I generally make the first feed after treatment a thinner and more watery consistency. The feeding spoon is washed and placed in Milton™ (a proprietary brand of disinfectant for infants). At each feed the spoon is removed from the disinfectant and rinsed with clean water before feeding.

During the second or third day of life outside the egg I decrease the brooder temperature to 36°C and gradually reduce the temperature usually by one degree over a one or two day period. On the fifth day the temperature is set and maintained at 34°C until the ninth or tenth day when I decrease it to 33°C. Once the chick is two weeks old, it can usually produce enough heat to keep itself warm, provided that the ambient temperature is not less than 26°C. If the weather is cold, I maintain the chicks until feathered at between 25–30°C, dependant on age, size of youngster and number of chicks sharing the brooder. Feel your chicks to determine if they are cold.

Chicks that are too hot will usually pant and spread themselves, lying on the nest material with wings held away from the body.

Chicks that are cold will usually call even though recently fed and will be cold to touch. Remember that chicks often voice their opinions and may not be cold. Therefore it is wise to always touch and observe them before you increase the temperature.

Some chicks can wean early, particularly in lory species. I do not believe that it is in the bird's interest to start weaning before it would leave the nest naturally. Some aviculturists insist on the chick feeding itself as soon as possible to obviously avoid the work involved in handfeeding. I have also observed a young bird that eventually died because a 'lori-culturist' was too lazy or overworked to feed the bird properly. The chick was forced to feed itself at an early age and was just not capable of doing it effectively. The aviculturist concerned justified the system adopted, by saying that the bird was being stubborn and when it died, justified its death by saying, 'I thought there was something wrong with it.' There certainly was. It should never have been unfortunate enough to have hatched in this particular aviculturist's collection. Every person is an individual and so are birds, some adapt to weaning quickly, others take longer and need our care.

All parrots have two toes forward and two backwards of the tarsus to grip perches. When born, the inside back toe is always forward, but as the chick develops, usually at approximately 21–23 days of age the toe settles itself to the back. If this does not occur, help can be given. Place the toe in its correct position and tape it to the tarsus with a Band-aid™. After a week, take off the restraint and the chick's toes should be perfect.

Lories As Pets

Lories and lorikeets make absolutely wonderful pets. Handrearing young is the easiest way to start adjusting a bird to be a companion to humans.

Even during the weaning stage, give the bird comfort for at least an hour per day. Just holding, cuddling and talking to your pet comforts it and gives it confidence. When very young they will snuggle into your hand, close their eyes and go to sleep; this helps

The author with two handreared Black Lories.

to humanise the bird.

As the bird develops during these sessions I 'preen' the bird with my fingers. I lightly rub my finger over the ear which invariably stimulates the bird to open its beak and stretch its neck as if yawning. When the quills start opening I help rub the loose wax off the bird as if I was a mate or parent. If this is done lightly the bird shows contentment and will even give you a turn by reciprocating the preening.

Talking to your bird will help it to become a talker. Not all lorikeets become talkers, but I have known some very capable mimics, especially in the genus *Lorius*. The larger species are generally the most adaptable to talking, but all can become very tame and friendly. A Chattering Lory owned by a friend in South Africa used to repeat sentences mimicking the voices of both his master and mistresses. He was so accomplished that he often fooled his human friends into thinking that they were actually calling each other and not the bird. I have since kept a Chattering Lory cock that has also amused me with his mimicking ability.

Lories and Lorikeets make wonderful pets like this handreared Red-collared Lorikeet.

Just how friendly, talkative and playful a pet bird becomes depends on the character of the bird and most importantly on how much time you are prepared to devote to it. Those people that put time into developing a pet will reap the rewards.

Some birds are adept at performing tricks and communicating, however make sure that your pet is closely involved in your day-to-day activities at home. The bonded bird will only suffer if you exclude it from these activities. Naturally, you need time to yourself, however keeping and maintaining a pet also means devoting a reasonable amount of your time to your pet. If you cannot, DO NOT keep a bird as a pet.

Obtain a suitable cage from a pet shop that allows the bird adequate exercise when confined. Always give the bird access to you as much as possible and depending on circumstances, allow the bird time out of its cage.

Birds love to fly around a room and then land on you. Lorikeets love to clamber about on your clothes hanging upside down. Swinging them on a rope or towel or something they can grip on to gives them great delight. Be gentle when you do play with them. They are a lot smaller than you. I can grab my pets by the feet and swing them to and fro which they love. Do not forget that sometimes they just love to siesta on your shoulder and cuddle into your neck. Provide toys or swings for your pet's amusement when you are busy and cannot attend to them. These important enrichment activities will enhance the overall physiological well-being of your bird. When your pet is having indoor supervised flying times out of its cage, the nature of its droppings needs to be considered. However a towel placed on your shoulder, over a chair or other items of furniture will assist in minimising the effect of the droppings.

It is much more noticeable when speaking to the pet shop trade that young, tame birds are in much demand and sell at a higher price. This is justified considering the amount of work necessary to create a tame bird. To assist in achieving a strong bond with your pet bird, it is suggested that the studying of book titles and videos dealing with the taming and training of birds be consulted. **ABK Publications** produce quite an excellent title, ***A Guide to Pet and Companion Birds***, which is highly recommended.

DISEASES AND DISORDERS

Since lories and lorikeets require feeding on a daily basis, the owner is able to observe the birds regularly. It is important to become familiar with your birds. Once you are familiar with an individual bird's behaviour and condition, anything abnormal should be noticeable within a 24 hour period. Since I feed twice daily, the birds' condition can be assessed even more frequently. At this point I highly recommend that **A Guide to Basic Health and Disease in Birds** (Revised Edition) by Dr Michael J Cannon, published by **ABK Publications** be part of your necessary resource material.

If your bird is fluffed up, its head is placed behind a wing with two feet on the perch, or if it is generally listless, has dull or closed eyes, a dirty vent and nasal discharge, be aware that either one or a combination of these factors would indicate an unhealthy bird.

It is vitally important to isolate this bird immediately for its own benefit as well as any other birds in its proximity. It is important to minimise the stress caused in catching the bird, but this is obviously extremely difficult to achieve.

Place the bird in a quiet, secure and comfortable hospital cage. The bird may need immediate treatment in some cases, but it is usually wise to provide your bird with heat (temperature of 28–30°C) to reduce the stress the bird is experiencing just to maintain body temperature.

Experienced avian veterinarians should be consulted if possible. I have been fortunate enough to have developed complete confidence in the avian veterinarians in my area. I have left birds in their care and have considered myself lucky to have their professional assistance. They will help you keep your birds vigorous and healthy.

These days, specialist veterinarians have the equipment and the knowledge to rectify most diseases and disorders. They can take swabs from the crop or the faeces and determine the ailment much more accurately than an aviculturist can. Another method adopted these days is the investigation of organs using a fibre-optic laparoscope in an anaesthetised live bird. The condition of a bird's internal health can be determined while surgical sexing takes place.

Dr Danny Brown BVSc (Hons) BSc (Hons) was very kind to add his knowledge concerning the health care of lories and lorikeets as follows:

Typical signs of a sick bird in a Musk Lorikeet.

Lories and lorikeets, by virtue of their lifestyle, pose several problems with regard to diseases. These birds consume large amounts of foods which are prone to contamination and produce copious wastes which contaminate the environment. The majority of infections seen in these birds therefore tend to be gastrointestinal in origin and the faecal-oral route of transmission is most important.

Hygiene is potentially the single most important factor involved in disease control in this group of birds. Their almost wholly arboreal nature is a godsend and significantly reduces the potentially disastrous risks associated with floor contamination below these birds. The use of suspended cages has further improved management in this respect. Feed stations and aviary furniture (such as perches and nest facilities) therefore represent more significant sites of contamination than in other parrot species.

The following is a brief summary of the most significant diseases of lories and lorikeets as seen in this country. These birds may be afflicted by many other disorders but the diseases listed are those that almost every lorikeet fancier will encounter at some time.

Crop Infections

Crop infections are one of the commonest presentations of the sick lory or lorikeet. These birds are often dull, listless and are either seen to vomit sticky mucous and food or have copious food and mucous around the facial feathering. The crop may either be empty due to persistent vomiting or full and often spongy to feel. Despite being quite ill, these birds often continue to eat ravenously but often vomit immediately after. Crop infections are mainly due to bacterial or fungal causes.

Bacterial crop infections may occur as a result of inappropriate changes in normal bacteria, eg with excessive use of antifungal drugs (resulting in an increase in bacterial numbers in the absence of competition for space or nutrients) or excessive intake of bacteria (eg poor feeding hygiene). Fungal crop infections are similarly developed (eg yeast populations may increase rapidly during and after antibacterial therapy). It is therefore important to use the correct treatment to avoid exacerbation of the disease. Your avian veterinarian can assist by taking crop samples and identifying the organism involved by microscopic examination. Medication may then be needed to stop the vomiting, treat the initial organism and possibly to prevent secondary organisms.

Many aviculturists will routinely treat any vomiting lorikeet with antifungal drugs such as Nilstat™. Using medication without knowledge of the actual organism involved has its risks. As well as the problems outlined above, this has led to significant resistance in some populations of *Candida albicans* yeast and they will no longer respond to commonly available drugs. This means that you may now be limited to the use of more expensive and potentially more toxic drugs as an alternative.

If dead birds are examined, the crop may be filled with thick mucous and the crop lining may be thickened and may be raised into white areas called plaques.

Crop infections may start at the top of the gastrointestinal tract but will often infect areas lower in the gastrointestinal tract causing severe diarrhoea and may affect other organs such as the liver. If vomiting is excessive, some material may be accidentally inhaled resulting in respiratory infection. If adult birds are infected, they may easily pass on organisms to young being fed in the nest. Young birds have poorer immune systems and may succumb very quickly.

Most cases of bacterial and fungal crop infections can be reduced or avoided by following suitable protocols for the cleaning of feeding utensils and aviary furniture, therefore reducing the build-up of organisms in the environment.

Psittacosis (Chlamydophilosis)

This is a disease caused by an organism called *Chlamydophila psittaci*. It is a very common infectious disease that will often be carried by a bird with no clinical signs being observed. When these birds become stressed, eg as a result of another illness, transportation, moulting or breeding, the subsequent reduction in the immune response may allow the organism to grow. *Chlamydophila psittaci* is spread rapidly between birds by respiratory or nasal excretions, faeces or feather dust. Clinically affected birds may show signs including swollen watery eyes, sneezing, nasal discharge, green diarrhoea, depression, anorexia and sudden death. Birds of all ages may be affected.

Diagnosis is best made by your avian veterinarian on the basis of clinical signs and testing of appropriate smears for the presence of the organism. At post-mortem, you may see air sac infection, an enlarged spleen or an enlarged liver.

Treatment is time consuming. The drug doxycycline is considered to be the drug of choice and needs to be used continuously for a minimum of seven weeks. Birds may clinically improve quickly but if the medication is withdrawn too early relapses are common. A bird that has been treated for Psittacosis develops no immunity to the disease and can be infected repeatedly. The treatment is given either as a weekly injection or as a daily medication in the water or nectar. Refer to *Medication*.

Psittacosis is a zoonotic disease (ie it can infect humans). It can cause influenza-like symptoms in susceptible persons. If you suspect Psittacosis in your bird and you have these signs inform your doctor that you keep birds and appropriate tests may then be initiated.

Psittacine Circovirus (Psittacine Beak and Feather Disease, PBFD)

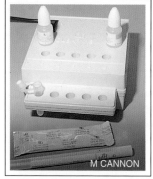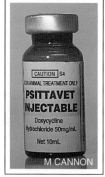

Above left: The Clearview™ test used to detect Chlamydophila psittaci (Psittacosis) in sick birds. It is not as useful in birds that are not sick.
Above right: Psittavet™ – the injectable form. One of the best antibiotics for Psittacosis.

This virus is known to affect both wild and captive lorikeets. It is caused by a circovirus and infects rapidly dividing cells within the body. This means that the cell population that will be most affected are the feather and skin cells which are constantly growing. Liver, spleen and bone marrow cells are also damaged by this virus.

Psittacine circovirus can cause very sudden illness and acute death but more commonly we will see the chronic form of the disease. The classical initial presentation is the loss of primary wing feathers and tail feathers. This is more common in younger birds. These infected birds are often referred to as 'runners'. The disease may later progress to further feather loss, secondary infectious problems associated with the liver, and spleen damage.

Lorikeets are one of the few parrot groups that show an apparent ability to 'recover' from this disease. Some birds will regrow all feathers and appear clinically normal. Research is still under way to determine if these birds are then free of the virus or if they are acting as lifelong carriers and will actively spread the virus. Until this research is completed and it is determined whether these birds are carriers, it is safest to consider these birds as infectious and cull them from your collection. The greatest risk is that the feather dust and faeces may remain infective in the environment for several months and are therefore a risk to the rest of the collection. It is still not confirmed whether another virus, polyomavirus, is also involved in this feather loss syndrome.

Many affected birds will invariably die of secondary complications and euthanasia of these birds may be a more responsible approach. A number of aviculturists have introduced this disease to their collection by offering to care for wild

Rainbow Lorikeet suffering from Psittacine circovirus.

'runner' lorikeets brought in by the well-meaning public.

Psittacine circovirus may be diagnosed definitively by blood and feather samples.

Prevention involves screening all incoming susceptible birds for the virus or buying your birds from circovirus-free collections.

If a bird in your collection is diagnosed with circovirus then you will need to disinfect the aviary with appropriate disinfectants (eg Avisafe™, Parvocide™) and discard any perches, nestboxes etc (which may act as a reservoir for virus trapped in feather dust) and replace the aviary floor substrate.

Loss of tail feathers due to Psittacine circovirus.

Parasitic Diseases

Parasitic worms are an uncommon cause of disease in these birds as they are more frequently kept in suspended aviaries and rarely utilise the ground in conventional aviaries. The exception to this is the occasional tapeworm infection. Tapeworm is acquired when a bird eats an insect host carrying the tapeworm larvae. Surprisingly, lorikeets do seek and eat moderate numbers of insects in their enclosure. Tapeworm may cause general illthrift and weight loss. Tapeworm segments may be seen in the droppings. Diagnosis is generally made by faecal examination for tapeworm eggs. Treatment is relatively easy using a drug called praziquantel (Droncit™, Virbac Tapewormer™, Wormout Gel™).

Coccidia is an uncommon protozoal parasite that causes diarrhoea, often bloody, as a result of intestinal lining damage. Coccidiosis is diagnosed by faecal examination of egg-like structures called oocysts. There are many treatments available but the most efficient is a drug called toltrazuril (Baycox™).

Trichomoniasis is caused by another protozoan parasite *Trichomonas gallinae* that lives in the crop and intestine. It may cause vomiting and diarrhoea. It can be spread via contact with contaminated water or nectar. It is best diagnosed by crop wash and examination of the mobile organisms. Treatment requires the use of drugs such as ronidazole or metronidazole.

Medication

In the event of an illness in a lorikeet, medication may need to be administered. Lorikeets are perhaps a little simpler than other species to medicate by virtue of their diet. The feeding of nectar to lorikeets provides for a simple route of medication. Most medications can be added to the daily nectar, which will often mask otherwise distasteful oral products. One problem associated with medicated nectar is that a lorikeet may consume, in a single day, much more nectar than an equivalent sized seed-eating parrot would consume in medicated water. For this reason, dose rates designed for in-water medication may need to be modified in order to avoid potential toxicity. For example, an average 100 gram seed-eating parrot will consume in an average day 6ml of medicated water. A 100 gram lorikeet may easily consume 30ml or more of medicated nectar. Depending on the drug, the dose rate may need to be at least one fifth the dose to achieve the same therapeutic effect. Your avian veterinarian will advise you when dose modifications are necessary.

Lorikeets can also be medicated easily (with some drugs) by direct oral means as they will readily lick any sweetened liquid from a syringe or dropper. Mixing medication with honey or sugar water may improve palatability.

Crop administration by crop needle can also be used. An average small lorikeet will safely accept 0.5–1.0ml and larger lories may accept 2–5ml crop doses. Always err on the side of safety and use the lower dose until you are comfortable with the technique.

Injectable medications are best administered into the breast (pectoral) muscle mass. This is an important medication method and may be essential if the bird is vomiting and therefore cannot tolerate oral medication. Your veterinarian can demonstrate and train you in these techniques.

INTRODUCTION TO
LORY & LORIKEET SPECIES

In this section, individual species of lories and lorikeets will be discussed. Those listed are known to exist in aviculture in Australia. Some species may exist that I have not been informed of and these have therefore not been included.

This section deals with specific information about individual species. You will need to consult earlier sections for information of a general nature about lories and lorikeets.

The family size range lies between the large *Chalcopsitta* species measuring approximately 32cm to the small lorikeets of the *Charmosyna* species namely the Wilhelmina's Lorikeet *C. wilhelmina* measuring only 13cm in length.

LORY AND LORIKEET SPECIES

Red-collared Lorikeet

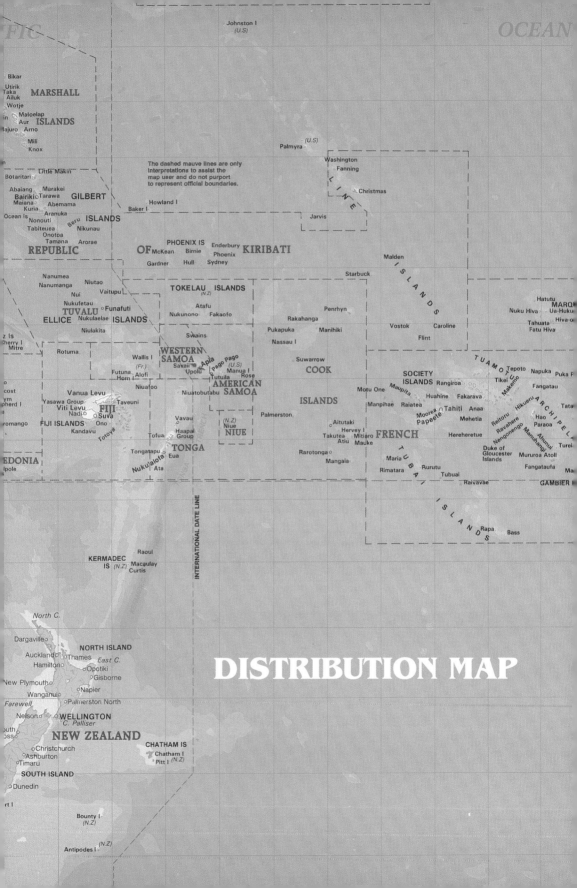

CHALCOPSITTA GENUS

Chalcopsitta = bronze or copper and parrot.

BLACK LORY

BLACK LORY
Chalcopsitta atra atra (Scopoli)
atra = black.
bernsteini = named after Bernstein.
insignis = extraordinary, remarkable, distinguished.

In the Wild
Distribution
Western New Guinea – Vogelkop and the islands of Salawati and Batanta.

Subspecies
The Berstein's Lory *C.a. bernsteini* is found on the island of Misool. Forehead and thighs variably marked red.

The Rajah Lory *C.a. insignis* is found on Onin and Bomberai Peninsulas, eastern Vogelkop and Amberpon Island.

Habits
This species is found in open savanna where small clumps of eucalypts are dispersed over the lowlands, along forest edges and in mangroves. Coastal coconut plantations and swamp forest are also favoured and they have been recorded feeding on the flowers of *Schefflera* scrubs. The birds are usually found in groups or small flocks of about ten birds, but occasionally form into larger flocks. The flight is direct but is not swift and consists of rapid shallow wing beats.

Breeding
Two white eggs are laid in a hollow of a tree. The nesting season (probably the same as many of the New Guinea parrots) tends to be year round with greatest activity between March and November.

Only in captivity was courtship display recorded by Bernard C. Sayers (*Aviculture Magazine*, UK 1974). 'One bird bounded sideways along a perch in typical lory fashion until about nine inches from the supposed female, both birds then stretched to their full height and indulged in short bursts of rapid wing-flapping interspersed by bouts of vertical bobbing. They then stretched their necks and turned their heads at right angles to the vertically held neck and although I could not hear through the glass aviary front, I assume that they 'sissed' as do other lories when indulging in a similar display. The supposed male then fed the female. The female then moved away, was followed by the male who mounted and copulated.'

Above:
Bernstein's Lory C.a. bernsteini.
Below:
Rajah Lory C.a. insignis cock bird. Note the unusual amount of red colouring – aberrant form.

Description

Cock

Length: Approximately 32cm. Weight: Approximately 220 grams.

In sunlight the black colouration has a purplish tinge, and the black body feathers are fringed with a light grey tone. Rump is blue and underside of tail red and yellow. Beak and feet are black. Iris is reddish brown with a thin yellow inner ring around the pupil.

Black Lory hen at nestbox.

Young at 15 days of age.

Hen

Same as the cock, but usually smaller. Iris is brown.

Immatures

Variably marked with red feathers, particularly on the head. Iris is dark brown. Skin around eyes and base of lower mandible is white not black as in the parents. This skin colour is likely to aid the feeding of a young bird by the parents in a dark nest.

Young at 24 days of age.

Voice

The Black Lory is a very noisy species and has a grating-like screech. The intensity of the screech depends on the level of excitement of the bird, becoming louder when it feels intimidated or exuberant. A softer tone is noticed when the birds are in close contact. This is probably a communication call between members of a pair and no doubt their offspring when feeding in close proximity to each other.

Aviculture

The Black Lory became available to aviculture in significant numbers during the early 1970s and has been bred on a number of occasions in South Africa, Europe, UK and the USA. Unfortunately, probably due to its lack

Young at 34 days of age.

of spectacular colours, it has never been popular. This is regrettable as they are extremely intelligent and interesting birds and in direct sunlight are extremely beautiful with their purple, black, blue and greyish tinges.

Young at 52 days of age.

Housing
In my experience the Black Lory requires a suspended aviary measuring at least 3.6 metres long x 90cm wide x 1.2 metres high. They have been bred in smaller enclosures but these are large birds and I cannot help but feel that in the long-term, it is better for the birds' health to be housed in a larger area. I house my birds in aviaries measuring 1.8 metres wide.

They are not fussy eaters and thrive on mixtures as described in the section on *Feeding*.

Black Lory hen left, with two recently fledged young.

Breeding
A nestbox measuring 30cm square x 60–90cm high with an entrance hole 10cm in diameter would prove sufficient. Black Lories have been known to become extremely aggressive towards their owners when breeding. Therefore it is suggested to minimise husbandry around a breeding pair. However, certain pairs may be tolerant of interference and accepting of their owners allowing minimal observation of the nesting procedure.

Clutch size is two eggs. Incubation lasts for approximately 24 days. Young fledge at approximately 75 days of age. In some cases two broods can be expected from a good breeding pair especially if young are taken for handrearing.

Mutations
There are no known mutations in this species.

DUYVENBODE'S LORY

DUYVENBODE'S LORY
Chalcopsitta duivenbodei duivenbodei **(Dubois)**
duivenbodei = named after Duyvenbode.
syringanuchalis = lilac nape of neck.

In the Wild
Distribution
Northern New Guinea between Geelvink Bay and Astrolabe Bay.

Subspecies
C. d. syringanuchalis is found in the eastern part of its range. It is doubtful as to the validity of this subspecies but it is considered to be darker on the head and back and has a dark violet sheen.

Habits
The Duyvenbode's Lory is found in lowlands, frequenting forests and tall secondary growth. This species is considered to be uncommon and is found in loose flocks of up to ten birds made up of pairs and no doubt juveniles. Birds roost socially. Flight is the same as for the Black Lory. The yellow underwings are particularly noticeable in flight. R. Low (1977) records what I can only describe as a threat display which is no doubt used to usurp competitors on flowering trees and in other aggressive circumstances.

She describes the display: 'Here they will lunge the head forward, uttering the typical complaining, almost growling sound, shared with the Black Lory. As they lunge forward, the tail is fanned, the shoulders or wings are lifted to reveal the yellow undersides – in fact, all the yellow parts of the plumage are thrown into prominence.'

Breeding
It is assumed that the birds lay two white eggs in a hollow of a tree as there are no records of nests having been encountered.

Breeding would take place at a similar time to most New Guinea lories, during the period March to November. In captivity, courtship display was recorded by R. Low (1977): 'The male revolves around the perch and then flaps the wings vigorously to reveal their most startling feature – the golden undersides – when hanging below the perch.'

Description
Cock
Length: Approximately 31cm.
Weight: Approximately 230 grams.
Predominantly dark brown with forehead, throat, bend of wing, underwing coverts and thighs a bright yellow. Yellow-white streaking appears on the nape and hindneck. Rump and undertail coverts are violet-blue. Beak and feet are black. Iris is dark brown with a thin yellow inner ring around the pupil.

Duyvenbode's Lory

Hen
Same as the cock but usually smaller and less bold in attitude. Colours are usually less pronounced.

Immatures
The yellow area of the face is less extensive and plumage duller than in the adults. Yellow streaking on nape and hindneck as in adults is absent. Iris is dark brown. The skin around the eyes and base of the lower mandible is white and not black as in adults. Bill and cere are black. Young are noticeably smaller than adults.

Voice
As for the Black Lory.

Aviculture
This is another of the New Guinea lory species which became available in reasonable numbers in 1973. In the UK, at the time of writing, they were selling for as low as $AUD200 each due to apparent lack of demand. I sincerely hope that English aviculturists do not let this beautiful species slip back to being a rarity, as I understand that the UK has a very prolific gene pool in comparison to other countries.

Housing
Minimum aviary size is as for the Black Lory. I have not had the opportunity to see this species in large aviaries, but I would like to see them displayed in an aviary measuring 6 metres long x 1.2 metres wide or larger. The birds would certainly display their lovely wing pattern in this environment.

Above: Pair of Duyvenbode's Lories. Right: Duyvenbode's Lory chick at 8 days of age.

Breeding
Nestbox size and husbandry are as for the Black Lory. This species has proved a reasonably regular breeder in Europe, the UK and the USA. Clutch size is two eggs with an incubation period of approximately 24 days. Young fledge at approximately 70–77 days of age. R. Low considers them to be conservative eaters.

Mutations
There are no known mutations in this species.

YELLOW-STREAKED LORY

YELLOW-STREAKED LORY
Chalcopsitta scintillata scintillata (Temminck)
scintillata = sparkle, clean, scintillate.
chloroptera = with green wings.
rubrifrons = with red forehead.

In the Wild
Distribution
Southern New Guinea from Kamrau Bay, Irian Jaya to Kemp River, south-east Papua New Guinea, and the Aru Islands.

Subspecies
C.s. chloroptera lacks red underwing coverts which are green in this eastern subspecies. This subspecies is found as far west as the upper Fly and Noord Rivers.

C.s. rubrifrons tends to have orange streaking on the breast as opposed to the yellow of the nominate species. This subspecies is found on the Aru Islands.

Habits
Found in lowland forest and adjoining open savanna. This species has been recorded in rainforest and dense bush along water courses near Port Moresby, Papua New Guinea. If you are familiar with the call of the *Chalcopsitta* species, they can be recognised clearly before being seen. I saw them on July 7, 1990 in open forest, west of Port Moresby where ten were seen feeding in a high tree, after I was made aware of their presence by their noisy characteristic calls. I was unable to identify the species of tree they were feeding on.

They have a direct flapping flight and are usually found in small noisy flocks.

H.L. Bell (Coates 1985) reports seeing them and other species of lory and honeyeater in a drunken condition. This condition apparently arose from the birds' partaking of fermented coconut juice which was being collected by villagers from green coconuts.

Breeding
A nest of this species was found by Mackay (in *Observations for September New Guinea Bird Society Newsletter* 1971) at the Veimauri River in a hollow approximately 24 metres above ground level. We presume that, as in captivity, the birds lay two oval white eggs on the rotten wood inside a hollow.

They are often seen hanging upside down acrobatically and performing antics. They hang by one foot with wings held open. These antics were recorded by B. Coates (1985) and are often seen in other species of lories and lorikeets. A pair were observed feeding well-grown fledglings near Brown River in early February by H.L. Bell (Coates 1985).

Pair of Yellow-streaked Lories.

Description

Cock
Length: Approximately 29cm. Weight: Approximately 200 grams.

Predominantly green on the back and wings with yellow to orange-yellow streaking on the head and breast. Forehead is red and there is red streaking on the nape, throat and upper breast. Back and feet are black. The underside of the wing has a yellow band which is quite pronounced against the red underwing coverts. *C. s. chloroptera* has green underwing coverts. Iris is orange-red.

Hen
Difficult to distinguish from the cock. Generally less intense in colouration, although this is not always the case. Surgical sexing is needed for accurate determination.

Immatures
Duller in colour than the adults. Iris is dark brown with the skin around the eyes and base of lower mandible white when recently fledged.

Voice
As for the Black Lory.

Aviculture

The early 1970s saw significant numbers being made available to aviculture particularly in Europe, the UK and South Africa. This species seemed to take a long time to adapt to captivity and begin breeding, even though the birds appeared content when being well maintained and accommodated. In fact this seemed to be generally the case with all *Chalcopsitta* species except the Cardinal Lory.

Housing
As for the Black Lory.

Breeding
As for the Black Lory. Clutch size is two eggs. Incubation lasts for approximately 24 days. Young fledge at approximately 70–77 days of age. It has been bred on many occasions but cannot be considered established.

Mutations
There are no known mutations in this species.

Yellow-streaked Lory C.s. chloroptera.

CARDINAL LORY

CARDINAL LORY
***Chalcopsitta cardinalis* (G.R. Gray)**
***cardinalis* = of the colour of a cardinal's cassock; deep scarlet.**

In the Wild
Distribution
This species is found on the Solomon Islands, the eastern Bismarck Archipelago, Bougainville, Buka, Nissan and islands off New Ireland namely Tanga, Feni, Tabar, and is said to be found on Lihir and New Hanover.

I have been to Lihir and a nearby island, Masahet. The Cardinal Lory is not present on Lihir. I spent three weeks on this island and their absence was also confirmed by Dr Ian Burrows (*Report for Lihir Gold Mine*) while carrying out a survey on the Common Scrub Fowl. He notes that their absence 'was surprising in view of its abundance around the coconut fringed airstrip of Masahet Island, which lies some ten kilometres north of Lihir'. I am sure that he is mistaken about sighting these birds on Masahet, which has never had an airstrip. This was confirmed by an expatriate Australian who had lived there for over 13 years and was married to a local lady. I am now wondering at the validity of sightings all the way to the north to New Hanover.

Subspecies
None recorded.

Habits
Found in primary and secondary forest in lowlands and adjacent hill country. Also recorded in coconut plantations. Considered numerous in most parts of its range. They are conspicuous with their loud calls. Usually seen in small flocks feeding on blossoms, preferring red flowering trees (Coates 1985). Large flocks traverse ocean between one island to another in search of food.

Breeding
Display is as for all *Chalcopsitta* species. These lories are often observed displaying with drooped wings which they slowly flap.

No breeding records are available, except of display seen in September by Cain and Galbraith (Forshaw 1973).

Description
Cock
Length: Approximately 31cm. Weight: Approximately 200 grams.

Predominantly red with red-brown on the back and wing. The naked skin at the base of the bill is speckled white, which is contrary to other *Chalcopsitta* species. Beak is red-orange and feet are black. Iris is red.

Hen
As for the cock.

Immatures
Duller in colour than the adults, with greyish yellow margins to the feathers of the underparts. The beak is black tending to orange-brown and the ear coverts are light brown to yellowish. Naked skin at the base of the beak is black.

Voice
As for the Black Lory.

Aviculture

This species has been exported from the Solomon Islands in reasonable numbers. Apparently, the Cardinal Lory is not a popular species with aviculturists in Europe. This seems extraordinary to me as they are colourful and full of character. We can only wish that this bird will become established in Australian aviaries.

Housing

As for the Black Lory.

Breeding

This species seems to settle and breed quickly in captivity. This may have some bearing on the fact that they are considered to be the most numerous of the *Chalcopsitta* species in the wild.

This species becomes particularly aggressive, possibly more so than other species of *Chalcopsitta*, when they begin breeding. They will attack you without fear and it may be necessary to take precautions with husbandry to avoid being bitten.

Clutch size is two eggs. Incubation lasts for approximately 24 days. Fledging occurs at approximately 70–75 days of age.

Mutations

There are no known mutations in this species.

EOS GENUS

Eos = the dawn, sunrise, the east – goddess of dawn.

BLACK-WINGED LORY

BLACK-WINGED LORY
Eos cyanogenia (Bonaparte)
cyanogenia = with blue cheeks.

In the Wild
Distribution
This species is found on Biak, Supiori and other islands in Geelvink Bay, West Irian, Indonesia.

Subspecies
None recorded.

Habits
Found in coastal forests and secondary woodland and considered by Bishop (1987 in Forshaw 1989) to be very conspicuous. Their brilliant red and black plumage is most spectacular as the birds pass swiftly overhead. Bishop also reported them to be common and widespread on Biak and Supiori Islands in June–July 1982 and April 1986. It is believed to be more difficult to locate this species near populated areas on Biak these days, no doubt due to trapping and habitat destruction. In November–December 1937 Ripley (Forshaw 1973) recorded it as common only near the shore in coconut plantations near Korrido village on Biak. It is still common as a pet on Biak.

Breeding
During June–July 1982 Bishop (1987 in Forshaw 1989) observed pairs in courtship on Biak and Supiori. A pair were also seen at a hole in a tree in primary forest.

Description
Cock
Length: Approximately 30cm. Weight: Approximately 160 grams.

Predominantly red with a deep violet-blue area on the side of the head around the coverts. Upper wing coverts and thighs are black. Yellow band on underside of primaries is seen particularly in flight. In my pair the cock had a more pronounced forehead and was slightly larger than the hen with a larger violet-blue ear patch area and larger beak. Beak is orange and feet are black. Iris is red with a white inner eye ring.

Hen
Similar to the cock.

Immatures
Duller in colour than the adults with feathers on underparts and neck heavily margined with greenish black. I have observed a youngster with blue on the head between the ear patches. Iris is brown and beak is black.

Voice
They have a high pitched screech and also utter gutteral sounds.

Immature Black-winged Lory chick.

Aviculture

The Black-winged Lory is my favourite species of the *Eos* genus and seems to be more often bred in the UK than in Europe. Some are kept in Australia and I can only hope that they breed sufficiently to establish the species.

Housing

This lovely species should be kept in an aviary with minimum dimensions of 3 metres long x 90cm wide x 1.2 metres high if suspended, or 1.8 metres high in a conventional aviary.

Breeding

I possessed two birds of this species which seemed particularly interested in each other. The first clutch yielded three infertile eggs which made me wonder if I had two hens, although the two birds were seen to copulate. The next clutch also had three eggs which were infertile. I was then fortunate enough to obtain a cock which was placed in the aviary with the two hens. It was hoped that the cock would choose the hen which most interested him

Black-winged Lory chick at 10 weeks of age.

or vice versa, as the hen may have done the decision making. Both hens immediately attacked the cock. When introducing a bird to new territory you should always observe your birds until you are satisfied that no harm will be done. This is especially necessary when two birds already govern that territory. If I had not been there I am sure that the cock would have been killed by the two hens. I then decided to remove one of the hens. Within two months the new human-matched pair had young.

Nestbox suitability depends on the pair. I suggest placing a box measuring 25cm square x 40cm high as well as a box measuring 25cm square x 50cm long at an angle of 30–45 degrees to the horizontal. The birds can then select the box which most satisfies them.

I have found that this species becomes extremely aggressive during breeding. Therefore care needs to be taken to avoid disturbing the birds off the nest.

Clutch size is two eggs with an incubation period of approximately 24 days. Young fledge at approximately 70–75 days of age.

Mutations

There are no known mutations in this species.

BLUE-STREAKED LORY

BLUE-STREAKED LORY
Eos reticulata **(S. Muller)**
reticulata = marked so as to resemble a net or network, reticulated.

In the Wild
Distribution
Tanimbar Islands, Indonesia, as well as Kai and Damar Islands where it was introduced.

Subspecies
None recorded.

Habits
Said to be rarer on its native Tanimbar Islands since habitat destruction and trapping have occurred in recent years. (The Goffin's Cockatoo was listed as a CITES Appendix I bird for these reasons.) It seems, however, that the Blue-streaked Lory may not be as endangered as was previously thought.

In October 1981 this species was found to be very common along the coast of Yamdena in the Tanimbar Islands. It was found in mangroves, plantations and secondary forest. It was also found inland in primary forest by Smiet (1985).

Breeding
It is presumed that this species lays two eggs in a decayed wood hollow in a tree. No records are available.

Description
Cock
Length: Approximately 31cm. Weight: Approximately 160 grams.

Predominantly red on head and underparts. The red thighs and upper and lower tail coverts are marked with black. Primaries, secondaries and greater wing coverts are variably marked black. Back and rump are red. Blue streaks feature on the ear coverts, nape and mantle. Beak is orange and feet are grey. Iris is red.

Hen
Similar to the cock, but usually lacks the bolder character and pronounced forehead bulge.

Immatures
Duller in colour than the adults, with underparts broadly edged with blue-black and the markings on the wings are less defined. Peri-orbital skin is whitish. Iris is brown. Beak is brownish.

Voice
A shrill screech and high pitch whistling.

Aviculture
This species was readily available in the UK and Europe from the early

Blue-streaked Lory.

1970s. However, I am concerned that this species may become endangered in the future, since wild trapped birds should become impossible to obtain to supplement captive held birds. We must not lose a species because it becomes readily available. Unfortunately, aviculturists tend to become complacent with a species. Suddenly breeding pairs are lost and it becomes difficult to obtain suitable replacements and a reliable source of young aviary bred stock is lost to aviculture.

This species is extremely attractive and easy to cater for.

Housing

As for the Black-winged Lory.

Breeding

The nestbox is the same as for the Black-winged Lory. Apparently some parents are noticeable feather pluckers of their young and varying diets have not necessarily prevented this from happening. Perhaps this is stress related and the birds should be left as undisturbed as possible. This however may not resolve the problem and it may be necessary to remove the young for handrearing. The parents may be bored, as in the wild, they would be foraging for food and not have time to pluck. In captivity, everything is on tap and naturally a lot of time is spent with the young. Perhaps we need to introduce more activity to the aviary in the form of branches etc to amuse the adult birds.

Clutch size is two eggs. Incubation lasts for approximately 25 days. Fledging occurs at approximately 84 days of age.

Mutations

There are no known mutations in this species.

RED LORY

RED LORY
Eos bornea bornea (Linné)
bornea = of Borneo.
cyanonothus = dark blue and spurious or counterfeit.
rothschildi = named after Rothschild.
bernsteini = named after Bernstein.

In the Wild
Distribution
Amboina, Saparua, Ceram, Buru, Goram, Ceramlaut, Watabela and Kai Islands, Indonesia.

Subspecies
Difficult to ascertain except in the Red (Buru) Lory *E.b. cyanonothus* which is smaller and has darker red plumage than the nominate species, the Red (Moluccan) Lory *E.b. bornea*. *E.b. rothschildi* is smaller and *E.b. bernsteini* is larger than *E.b. bornea*. In Australia, the dark Buru subspecies is now available thanks to the limited time of permitted importation.

Red (Buru) Lory E.b. cyanonothus.

Habits
This species is very common in plantations near villages and surrounding secondary forest throughout the coastal areas of Amboina, Ceram, Buru and the Kai Islands. They are scarcer inland away from the coast. Large flocks foraging in flowering trees have been recorded. On moonlit nights, large roosting flocks often fly and screech much as Galahs are noted to do in Australia. Stomach contents have shown fragments of flowers and the remains of small insects.

Breeding
Stresemann (Forshaw 1973) saw several young birds taken from nests in hollows high up in old trees, in mid-December (Forshaw 1973). Adults hiss and arch their bodies over the top of the partner, and wing-whirring is often performed. The cock is usually more active in his efforts to seduce the hen to copulate.

Description
Cock
Length: Approximately 31cm. Weight: Approximately 170 grams.
Predominantly red with blue and black areas on the wings. Undertail coverts and vent are blue. Beak is orange and feet are grey. Iris is red.

Hen
Similar to the cock but usually noticeably smaller. The head of the cock is broader.

Immatures

Duller in colour than the adults. Young often have more blue on the cheek, thigh, breast and abdomen areas. This blue colour is lost at the first moult. Iris is greyish. Beak is brownish.

Voice

A shrill screech.

Aviculture

Commonly bred in captivity with Australia apparently producing an unusually high percentage of hens. With the last importation of Red Lories it is hoped that this imbalance will be a thing of the past. This species is noisy, but is usually acceptable to neighbours in typical urban situations. This species is brilliantly coloured and I enjoy seeing the admiration by Red Factor Canary breeders for its depth of red colouration. They are totally envious. If the birds are fed as prescribed in the section on *Feeding*, your birds will not lose their colour. Poor nutrition can fade the intensity of red colour.

Immature Red Lory.

It seems that Red Lories also have an abnormally high number of feather pluckers in comparison to other species held in captivity. There are many reasons suggested for this plucking syndrome. However, I have never been able to propose a definite solution but feel that this is a stress related syndrome.

Some pairs may benefit by changed conditions, such as a new aviary located in a new environment. Adjacent birds, although they may not have direct access to a feather plucker, may still be causing it stress.

A similar diet can see your entire collection including other Red Lories in perfect feather, with still one or a few feather pluckers.

Housing and Breeding

As for the Black-winged Lory.

Mutations

There are no known mutations in this species.

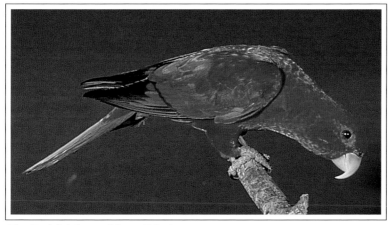

The Red (Moluccan) Lory E.b. bornea.

RED AND BLUE LORY

RED AND BLUE LORY
Eos histrio histrio (P.L.S. Muller)
histrio = stage player, ie dressed in colours like a harlequin.
challengeri = for HMS 'Challenger'.
talautensis = of the Talaud Islands.

In the Wild
Distribution
Sangihe, Talaud and Nenusa Islands between Sulawesi (Celebes) and the Philippines, which are part of Indonesia.

Subspecies
Recent reports suggest that the nominate species may be close to extinction due to loss of habitat. *E.h. talautensis* is considered to have less black on the wing coverts and flight feathers. This subspecies has recently become available in captivity, through trapping on the Talaud Islands.

E.h. challengeri has the blue band on the chest variably mingled with red. The blue line running from the eyes to the blue mantle is not continuous as in the other two representatives. This subspecies comes from the Nenusa Islands.

Habits
This species has been affected by land clearance for coconut plantations particularly on Sangihe Island since the latter part of the 19th century. They had retreated into the mountainous interior. It was considered at the same time to be extremely common on the Talaud Island group, with large flocks flying between islands.

In May 1986, it was considered common in small groups. Pairs were seen feeding in flowering coconuts within 100 metres of a village on the south-east coast of Karakelong Island, Talaud group, by Bishop (1987 in Forshaw 1989). Local people keep individual birds as pets and

Red and Blue Lory.

Red and Blue Lory chick at 38 days of age.

confirmed the bird to be common throughout the island.

In recent years this species has become available to aviculture and concern has been placed on its long-term survival by being placed in the CITES I category. This bird seems to be more readily available now in Europe and South Africa than a few years ago. The numbers which have been taken for aviculture should see it being established in captivity. It is extremely important that aviculturists do not exert further pressure on the wild population. In my opinion the only further justification for birds being taken from the wild is to establish captive breeding programs for the nominate subspecies and of course *E. h. challengeri*.

They live on islands and land clearance has been the devastation of this species. When large tracts of land are cleared, I see no reason why a species in this particular area could not be caught for aviculture before deforestation. These captured birds if left in the wild would place extreme pressures on adjoining populations by colonisation after deforestation. If this is not considered, high mortality will occur not only in one restricted area, but will spread to these adjacent forested regions.

In 1995 Jon Riley from the UK established 'Action Sampiri' which involved an expedition of ornithologists visiting the Sangihe Islands with the Red and Blue Lory as the focal species. The expedition discovered that this lory survived on only one of the far islands of the Sangihe group from which it had been recorded.

Breeding

Undescribed in the wild, but it is assumed that two oval, white eggs are laid in a tree hollow.

Description
Cock

Length: Approximately 31cm. Weight: Approximately 160 grams.

Predominantly red with dark violet-blue breast. A blue band extends across the crown over the eye, through the ear coverts to the hindcrown and the mantle. Uppertail is reddish purple and undertail is red. Beak is orange and feet are grey. Iris is red.

Hen

Similar to the cock but usually smaller.

Immatures

The underparts are variably marked with bluish black. Overall colour is suffused with more areas of

Juvenile Red and Blue Lory.

blue or black. Young specimens have a black bill tending towards orange-yellow as age progresses.

Voice
A typical lorikeet screech.

Aviculture
Since this bird has recently been introduced to aviculture there have already been some breeding successes. This bird is very rare in Australia and it is hoped that in time the species can be established. With Australia's favourable climatic conditions, I feel that it is important for aviculture in this country to have the opportunity to support captive breeding programs. Husbandry is as for the Black-winged Lory.

Housing and Breeding
As for the Black-winged Lory.

Mutations
There are no known mutations in this species.

VIOLET-NECKED LORY

VIOLET-NECKED LORY
Eos squamata squamata (Boddaert)
squamata = scaly or scaled.
riciniata = veiled, it has a prominent violet neck collar.
atrocaerulea = black and blue.
obiensis = from the island of Obi.

In the Wild
Distribution
Western Papuan Islands in Irian Jaya as well as Maja, Weda and the northern Molucca group of islands.

Subspecies
The subspecies *E.s. riciniata*, found on Weda and the northern Moluccas, has a prominent violet-grey neck collar often extending up to the hindcrown. The validity of the subspecies *E.s. atrocaerulea* requires further investigation. A distinctive subspecies, *E.s. obiensis* exhibits very little blue and blue-grey feathering around the head, neck and chest, if any at all.

Habits
This species was observed flying between Haitlal and Misool, as well as feeding in coconut plantations. Birds were observed congregating in the early morning and later evening in the flowering crowns of *Erythrinia* trees. They have been recorded in mangroves, coastal forest and montane forest up to an altitude of 1000 metres on Morotai.

Breeding
No records available of breeding in the wild. It is presumed that they lay two white, oval eggs in a tree hollow.

Description
Cock
Length: Approximately 27cm. Weight: Approximately 110 grams.
Predominantly red with variable violet-blue collar. Underparts are violet-blue, uppertail is purple-red and undertail is brown-red. Beak is orange and feet are grey. Iris is yellow to red.

Hen
Similar to the cock. Usually the cock has a more pronounced forehead and is slightly larger and bolder in appearance.

Immatures
Duller in colour than the adults with the underparts suffused with purplish black. The blue on the ear coverts and crown is variably found in young. Recently fledged chicks have a

Violet-necked Lory E.s. riciniata.

white peri-orbital ring. Beak and iris are brown.

Voice

A shrill screech.

Aviculture

A desirable and small species compared to those available in aviculture from the *Eos* group. Its husbandry is the same as for the Black-winged Lory. Australia has a number of birds and I believe that at least one pair are regular breeders.

Housing and Breeding

As for the Black-winged Lory.

Mutations

There are no known mutations in this species.

Violet-necked Lory E.s. riciniata.

PSEUDEOS GENUS
Pseudeos = false Eos.

DUSKY LORY

DUSKY LORY
Pseudeos fuscata (Blyth)
fuscata = somewhat dusky.

In the Wild
Distribution
The entire New Guinea mainland as well as Salawati and Japen Islands.

Subspecies
Species are variable in colouration, therefore there is no defined reason for a subspecies. There are two colour phases, being Red-orange and Yellow. Breeding from a Yellow pair can produce Red phase birds.

Habits
An extremely social bird usually found in relatively large flocks which can be very noisy. If Dusky Lories are in your area you will be well aware of their presence without seeing them. I found them common in coconut plantations to the west of Port Moresby, Papua New

Above: Dusky Lory – Yellow phase.
Right: Dusky Lory – Red phase.

Guinea, on July 7, 1990, where they reacted to our presence with interest and much scolding. Whether they directed their calls at us as intruders into their domain, I am not sure, but the whole coconut plantation seemed to erupt into abuse by the large flock present when we alighted from our vehicle.

This species occurs up to an altitude of 2,400 metres, apparently preferring forest and partly cleared areas in the hills and lower mountains. They are also found in parks and gardens of towns throughout New Guinea. They seem to be highly nomadic with claims that the birds migrate from one side of the island to the other. In the Port Moresby region they have been observed in upper canopy trees of rainforest and its edges. They are relatively fearless when intruders are observing them. Dusky Lories have been recorded in immense flocks when travelling to feeding grounds or roost sites. Besides fruit, pollen and nectar they have also been observed feeding on moth pupae. Birds are often observed displaying with drooped wings and white rump feathers fluffed out, all the while shouting to their mate.

Breeding

In various parts of the Eastern Highlands, New Guinea records show that breeding takes place between November and April. Specimens with enlarged gonads were collected in early August in the same region and during July in the Arfak Mountains. In Bulolo, Morobe district on July 11, 1990 I observed a flock in the town with a youngster being fed. At the same time a young Goldie's Lorikeet was observed visiting the aviaries of Peter Clark in the same town. Nests have been recorded in hollows in tall montane trees. Nesting has also been reported at Veimauri River in September and well-grown fledglings were observed being fed by a pair at Brown River in early February (Coates 1985).

Voice

Very similar to *Chalcopsitta* species with slightly less intensity. This is an extremely noisy species and pairs will continue with a whine-like call prior to the breeding season. However, having said this, they are wonderful birds, full of character and very playful.

Description
Cock
Length: Approximately 25cm.
Weight: Approximately 155 grams.
There is a large variation in plumage colour in the phases. The cock is predominantly brown with a band of red-orange (Red phase) or yellow (Yellow phase) on the throat and upper breast and abdomen. Crown is either red-orange or yellow. The white rump is usually hidden by the wing primaries. The large beak is orange and feet are black. Iris is red to orange.

Hen
Similar to the cock but often less robust in its physical appearance, ie the cock has a larger and bolder head.

Immatures
Darker than the adults, with less defined markings. The skin around the base of the beak is grey. I have never noticed orange flecks in the peri-orbital skin until the chicks are much older. Bill and iris are brownish.

Fertile Dusky Lory eggs.

Dusky Lory chicks, seven and eight days of age.

Dusky Lory chicks, four and a half weeks of age.

Dusky Lory chicks, seven weeks of age.

Aviculture

Large numbers became available in Europe and the UK during the early 1970s. They proved to be prolific, so much so, that they are difficult to sell and have reduced in price so that many aviculturists can afford them. In Australia, on the other hand, numbers are still low. However it must be considered to be one of the more common exotic lories available. They are being regularly bred in Australia. On first acquaintance, the Dusky Lory is admired for its colours and antics more often than any other lory species.

Housing

The minimum sized aviary measures 3 metres long x 90cm wide x 1.2 metres high if suspended and higher for conventional aviaries.

Breeding

A nestbox 20cm square x 30–50cm deep will be acceptable. In my experience this species are seasonal breeders. Some breeders have found them prolific, but some pairs can be slow to the task. Also they can prove to be inconsistent in their breeding results.

Mutations

There are, as previously mentioned, two naturally occuring colour morphs, the Red and the Yellow. Whilst the Yellow colour morph is the more prevalent in nature, the Red colour morph is more likely to be the original wildtype and the Yellow colour morph a naturally occuring mutation.

TRICHOGLOSSUS GENUS
Trichoglossus = bristle tongued

RAINBOW LORIKEET

TRICHOGLOSSUS GENUS
Trichoglossus = **bristle tongued**

In the Wild
Subspecies
Of the 21 subspecies of the *Trichoglossus* genus listed in *Parrots of the World* (Forshaw 1989), 11 are kept and bred by Australian aviculturists. Three of these subspecies, the Rainbow *T.h. moluccanus*, the Red-collared *T.h. rubritorquis* and the Scaly-breasted *T.h. chlorolepidotus*, are endemic to Australia.

Habits
The *Trichoglossus* genus of generally lowland lorikeets are found up to an altitude of 1000 metres with a noticeable decline in abundance above this level. The Ornate Lorikeet is considered a common inhabitant of wooded mountain country up to about 1000 metres and avoids dense primary forest. It is seldom seen on the coastal flats. The Perfect Lorikeet on Timor, tends to be found at higher altitudes than the Rainbow Lorikeet and replaces it to 2300 metres. It was only recorded once below 1000 metres by Bruce (1974 in Forshaw 1989).

They are known to travel long distances to reach trees in blossom, at times covering reasonable distances over the sea between islands and the mainland of their distribution. Birds in suburban areas have even learned to accept food preparation from humans and indeed, during a drought in 1991, I fed large numbers in my garden. Initially, I estimated that we were feeding approximately 100 birds but after noting the number of times per day the birds were fed and the quantity of food provided by my wife, I decided to make a count. We were actually feeding between 300–400 birds per day which I must say was a bit of a drain on the pocket.

Display
The *Trichoglossus* species have rather amusing and elaborate displays which include walking along branches with an air of confidence, bobbing and strutting to each other. Wing-whirring often occurs as either a threat or a dominant form of display to hens and other cocks lower in the pecking order. This genus have powerful beaks and a bite can easily incapacitate an individual. Confrontations usually result in the less dominant bird retreating in haste. However I have seen birds, when feeding on the ground, roll on their backs in a submissive posture to pacify the dominant bird. Both cocks and hens display and they can join forces to oust an opponent or opponents in disputes over territory, food and nest sites.

Wild Rainbow Lorikeets feeding in the author's garden.

Voice

A shrill screech in all species. Many other vocalisations occur during courtship and pair behaviour. Unfortunately, it is extremely difficult to translate these various calls into words. It is also interesting to note that the calls of the *Trichoglossus* genus can vary between subspecies. For example, the Messena's Lorikeet *T.h. massena* has very similar vocalisations to those of the Australian *Trichoglossus* members, but some calls heard on Lihir Island, New Ireland Province, Papua New Guinea, were more strident than the *Trichoglossus* species found in south-east Queensland, Australia.

Breeding

The *Trichoglossus* genus generally lay two eggs in a hollow of a tree. Mary Le Croy, W.S. Peckover and Karol Kisokau recorded (in an unpublished paper) that the birds dig a nest chamber in the ground on the island of Poy-yai in the Admiralty Islands, Manus Province, Papua New Guinea. 'These burrows were horizontal and sloped slightly downward from ground level at the front.' Nests containing eggs were found on July 27 and July 28, 1988. According to Peckover (pers. comm.), he felt that the birds had adequate nest sites in surrounding trees but seemed to prefer the ground. The Ornate Lorikeet *T.h. ornatus* has been recorded as breeding in September and October. In Australia there is a period of peak breeding activity in the Rainbow and the Scaly-breasted Lorikeet which differs throughout their enormous range. However, these birds can breed throughout the year, although this would depend primarily on the abundance of food.

Young accompany their parents on feeding expeditions at an early age after leaving the nest and become independent very quickly. Most young in my region of south-east Queensland, Australia can be seen or heard (incessantly begging for food) during late July through to October. The species has been known to breed all year round under suitable conditions. Scaly-breasted Lorikeets share a very similar range as Rainbow Lorikeets, but are less dominant and give way to their bigger relatives in feeding disputes. Natural hybrids between Rainbow and Scaly-breasted Lorikeets do occur in the wild, however during my many years of observation I have only once found a hybrid. Incubation term is approximately 24 days, with youngsters fledging at about 55–60 days of age. Rainbow Lorikeets have been recorded breeding at one year of age.

Aviculture

This genus is very easy to cater for. They have been kept successfully on a seed diet and have bred under these conditions. All subspecies of the *Trichoglossus* genus are held in very low numbers in Australian collections, with the exception of the Australian species (the Rainbow, the Red-collared and the Scaly-breasted Lorikeet), the Ornate and the Perfect Lorikeet *T.h. euteles*. However, aviculture has progressed in leaps and bounds in lory nutrition. This combined with surgical sexing, will hopefully make these rarer forms become more readily available.

Housing

Birds can be housed in either suspended or conventional aviaries. Suspended aviaries should be a minimum size of 2 metres long x 90cm wide x 90cm high, although it is preferable to increase this to a length of at least 2.7 metres.

Breeding

Nesting facilities are varied and many nestbox designs are accepted by these species. In the wild it seems that most lorikeets prefer hollow logs which are horizontal or at an angle of about 45 degrees. This does not mean that they will not accept vertical boxes. The recommended size is 20cm square x 30–37.5cm deep.

For any reader wishing to gain further understanding in regards to subspecies recognition, I would suggest that he or she refer to A.J. Cain's *A Revision of Trichoglossus haematodus and of the Australian Platycercine Parrots*, Ibis 97.

RAINBOW LORIKEET
Trichoglossus haematodus moluccanus
haemotodus = resembling blood, blood red.
moluccanus = of the Moluccas.

In the Wild
Distribution
From Cape York Peninsula down the east coast of Australia to, and including, Tasmania. Isolated population, probably introduced or escapees in Perth, Western Australia. Also found from Bali through the Indonesian Islands, New Guinea, Solomon Islands, Vanuatu, New Caledonia and Loyalty Island.

Habits
In Australia, the Rainbow Lorikeet is found in all types of wooded country from the coast to as far inland as about 350km from the sea in Queensland. It is an extremely abundant species.

The Rainbow Lorikeet is a very sociable species which has been found in flocks in their thousands depending on conditions. However, when travelling in Australia it is certainly more usual to see small flocks of 10–20 birds, with greater concentrations at feeding or roosting sites. During the breeding season pairs tend to be more solitary but often form larger groups during feeding periods away from the nest. They are noisy, argumentative birds and a great deal of noise can be heard from flocks at feeding or roosting sites.

Description
Cock
Length: Approximately 30cm. Weight: Approximately 145 grams.

Head is violet-blue with blue-black shaft streaking. Abdomen is the same colour. The hindneck collar is yellowish green, breast is predominantly red marked irregularly with yellow. The rest of the underparts and back is predominantly bright green. Underwing coverts are orange-red with a broad yellow underwing band. Legs and feet are dark grey. Iris and beak are red to orange-red.

Rainbow Lorikeet chicks approximately four weeks of age.

Hen
Same as the cock, perhaps slightly duller in appearance.

Immatures
Generally duller in colour than the adults. Beak and iris brownish orange to dark brown.

Mutations

The Rainbow Lorikeet is currently riding a wave of new interest due to the appearance and establishment of a large number of new mutations. There are now sufficient mutations to provide numerous different combinations, adding even further to the range of colours being bred. Sadly, a Blue mutation has not reappeared and remains the one important 'gap' in the range of colours. The other area of contention has been the naming of different mutations. Already, breeders have learnt that the mutation originally known as 'Olive' is in fact correctly known as Greygreen. Unfortunately, many of the new mutations are also being named prematurely and renaming will become necessary in the future. The mutation commonly referred to as 'Cinnamon' in Australia is in fact a Dilute mutation and is quite distinct from the Cinnamon from Europe illustrated in the photographs. If breeders do not like the name Dilute, then the acceptable alternative would be to call them Pastel (the European term for Dilute mutations).

The following mutations have been, or are being, established in Rainbow Lorikeets: Greygreen, Dilute (incorrectly called Cinnamon in Australia), Melanistic (Blue-fronted), Lutino (sex-linked), Fallow (possibly more than one mutation), Recessive Pied, Cinnamon (Europe) and Blue (extinct). There are also two new mutations known as 'Olive' and 'Aqua', which are currently being developed. Both are probably misnamed but more study is required. Finally there are the so-called 'Acquired Yellow' birds which reportedly would not reproduce. There are, however, recent reports of successful breeding, so they deserve further investigation. This mutation would be a Black-eyed Yellow (Clear) if its establishment is proven.

1. Melanistic (Blue-fronted) mutation.
2. Cinnamon (European) mutation.
3. Blue (extinct) mutation.
4. Dilute ('Cinnamon') mutation.

1. 'Jade' mutation.
2. Black-eyed Clear (Yellow) mutation.
3. Dilute Greygreen (Mustard) combination.
4. Dilute ('Cinnamon') mutation.
5. Greygreen mutation.
6. Lutino mutation.

RED-COLLARED LORIKEET
Trichoglossus haematodus rubritorquis
rubritorquis = red-collared.

In the Wild
Distribution
Habitat extends across northern Australia from Broome, Western Australia, north and eastwards to the Gulf of Carpentaria, Queensland.

Description
Cock
Length: Approximately 30cm.
Weight: Approximately 130–150 grams.

Differs from the Rainbow Lorikeet in that the hindneck collar is bright orange, bordered below by a band of dark blue merging into blue and orange scaling over the lower hindneck. Remainder of the back is dark green. Breast is more yellow-orange and uniform than *T.h. moluccanus*. Abdomen is darker green-black. Legs and feet are a light grey-brown. Iris and beak are red to orange-red.

Hen
Same as the cock.

Immatures
Duller versions of the adults. Beak and iris are dark brown-black.

Red-collared chick 21 days of age.

Mutations

Mutations occurring in the Red-collared Lorikeet have been transferred from the Rainbow or Scaly-breasted Lorikeet through hybridisation and include the Greygreen ('Olive'), Dilute ('Cinnamon') and more recently the Lutino.

See *Mutations* page 143 and section on Rainbow Lorikeet mutations for explanation of names.

Above:
Greygreen Red-collared mutation.

Right:
Pied Red-collared mutation.

Below:
Lutino Red-collared mutation.

SCALY-BREASTED LORIKEET

Trichoglossus chlorolepidotus
chlorolepidotus = greenish yellow and scaly, covered with scales.

In the Wild
Distribution
Habitat extends from the east coast of Australia from as far north as Cooktown, Queensland to approximately the New South Wales/Victorian border in the south. An isolated population has been established around Melbourne, Victoria, probably from captive escapees.

Description
Cock
Length: Approximately 24cm.
Weight: Approximately 85 grams.
Medium-sized, bright green lorikeet. Head is green, tinged with blue on the crown, with scaly yellow markings on the lower hindneck,

mantle and more prominently on foreneck, breast and abdomen. Underwing coverts and the broad band across the underside of the flight feathers are bright red-orange which is unique among lorikeets. Legs and feet are grey. Iris and beak are orange-red.

Hen
Same as the cock.

Immatures
Duller in colour than the adults. Beak and iris are dark brown.

Above: Greygreen mutation.
Right: Blue mutation.

Mutations

Mutations established or recorded for the Scaly-breasted Lorikeet include the Greygreen ('Olive'), Dilute ('Cinnamon'), Blue and Lutino.

See *Mutations* page 143 and section on Rainbow Lorikeet mutations for explanation of names.

Above:
Dilute Greygreen (Mustard) combination (left),
Dilute ('Cinnamon') mutation (right).

Right:
Lutino mutation.

ORNATE LORIKEET
Trichoglossus ornatus
ornatus = decorated, adorned.

In the Wild
Distribution
The Ornate Lorikeet is found on Sulawesi (Celebes) and most larger offshore islands. The Ornate Lorikeet is considered a common inhabitant of wooded mountain country up to about 1000 metres in altitude and avoids dense primary forest. It is seldom seen on the coastal flats.

Description
Cock
Length: Approximately 25cm.
Weight: Approximately 95 grams.
Forehead, crown and ear coverts are dark blue with a patch of yellow behind ear coverts. Cheeks, throat and side of neck are red. Breast is barred with red and dark blue. Abdomen and back are dark green with variable yellow markings on the lower breast. Underwing coverts are yellow. Beak is orange. Iris is orange-red.

Hen
Similar to the cock.

Immatures
Less barring on the breast and more yellow on the abdomen. Iris and beak are dark brown.

Mutations
There are no known mutations in this species.

GREEN-NAPED LORIKEET

Trichoglossus haematodus haematodus haematodus = resembling blood, blood red.

In the Wild
Distribution
Habitat extends from western New Guinea along the north coast, east to Humboldt Bay and south to the Upper Fly River and includes the islands of Western Papua and Geelrink Bay, Amboina, Buru, Ceram, Ceramlaut, Goram and Watubela.

Description
Cock
Length: Approximately 26cm. Weight: Approximately 130 grams.

Head is blue to blue-black including ear coverts and throat. Nape band is a greenish yellow. Breast is red with purple-black barring. Abdomen is dark green. Thighs and undertail coverts are yellow edged with green. Underwing coverts are red, with a broad yellow band across underside of flight feathers. Back, wings and uppertail feathers are green. Beak is orange-red. Iris is red.

Hen
Similar to the cock.

Immatures
Duller versions of the adults. Beak and iris are dark brown-black.

Mutations
There are no known mutations in this species.

Green-naped Lorikeet chicks, artificially incubated and handreared, at 31 and 33 days of age.

MITCHELL'S LORIKEET
Trichoglossus haematodus mitchellii
mitchellii = named after Mitchell.

In the Wild
Distribution
Bali and Lombok, Indonesia.

Description
Cock
Length: Approximately 23cm.
Weight: Approximately 100 grams.
Head, cheeks and throat are dark brown-green, with a yellow-green nape band. Breast is scarlet red, abdomen is blue-black. Back and wings are dark green. Beak is orange. Iris is orange-red.

Hen
Hens are said to have a yellow fringe on their red breast feathers making their breast colour yellowish orange.

Immatures
Duller versions of the adults. Beak and iris are dark brown-black.

Mutations
There are no known mutations in this species.

WEBER'S LORIKEET
Trichoglossus haematodus weberi
weberi = named after Weber.

In the Wild
Distribution
Inhabits rainforests and casuarina stands on the island of Flores in Indonesia.

Description
Cock
Length: Approximately 23cm.
Weight: Approximately 85 grams.
Overall plumage is of various shades of green. Breast is greenish yellow as are nape collar and underwing coverts. Beak is orange. Iris is orange-red.

Hen
Similar to the cock.

Immatures
Duller versions of the adults. Beak and iris are dark brown.

Mutations
There are no known mutations in this species.

EDWARD'S LORIKEET
Trichoglossus haematodus capistratus
capistratus = provided with a mask.

In the Wild
Distribution
Inhabits the island of Timor.

Description
Cock
Length: Approximately 27cm.
Weight: Approximately 130 grams.

Forehead, cheeks and chin are dark blue merging with dark green shaft streaks around eye, ear coverts and lower throat. Nape collar is yellow as is the upper breast. Abdomen, wings and back are dark green. Thighs are dark green and yellow. Underwing coverts are yellow with orange flecks. Beak is orange. Iris is orange-red.

Hen
The yellow upper breast feathers in the hen are usually edged with green. The cock shows red edged feathers.

Immatures
Duller versions of the adults. Beak and iris are dark brown.

Mutations
There are no known mutations in this species.

ROSENBERG'S LORIKEET
Trichoglossus haematadous rosenbergii
rosenbergii = named after Rosenberg.

In the Wild
Distribution
Habitat is the rainforest of Biak Island in Geelvink Bay, New Guinea.

Description
Cock
Length: Approximately 28cm.
Weight: Approximately 130 grams.

Head is dark blue with a wide yellow nape collar. Breast is red with heavy blue-black barring. Abdomen is dark blue. Underside of the flight feathers has a broad orange band. Beak is orange. Iris is orange-red.

Hen
Similar to the cock.

Immatures
Duller versions of the adults. Beak and iris are dark brown.

Mutations
There are no known mutations in this species.

MASSENA'S LORIKEET
Trichoglossus haematodus massena
massena = named after Massena.

In the Wild
Distribution
Inhabits the Bismarck Archipelago of New Guinea, the Solomon Islands and the New Hebrides. Found in coconut plantations and frequents human populated areas.

Description
Cock
Length: Approximately 25cm.
Weight: Approximately 120 grams.
Forehead is dark blue merging to blue-black at nape, ear coverts and neck. Nape collar is green. Breast is red with thin blue-black barring. Abdomen, back and wings are dark green. Beak is orange. Iris is orange-red.

Hen
Similar to the cock.

Immatures
Duller versions of the adults. Beak and iris are dark brown.

Mutations
There are no known mutations in this species.

PERFECT LORIKEET
Trichoglossus euteles
euteles = perfect, without blemish.

In the Wild
Distribution
Timor and the Lesser Sunda Islands in Indonesia. Tends to inhabit higher altitudes, between 1000 metres and 2300 metres.

Description
Cock
Length: Approximately 25cm.
Weight: Approximately 95 grams.
Plumage is predominantly yellow and green without distinctive barring. Head is olive-yellow with light green nape collar. Breast and abdomen are yellowish green. Back, wings and tail are green. Beak and iris are orange-red.

Hen
Appears to be more greenish than the cock and shows less yellow on the head.

Immatures
Duller versions of the adults. Beak and iris are dark brown.

Mutations
There are no known mutations in this species.

PSITTEUTELES GENUS

psitteuteles = parrot and perfect, without blemish.

GOLDIE'S LORIKEET

GOLDIE'S LORIKEET
Psitteuteles goldiei (Sharpe)
goldiei = named after Goldie.

In the Wild
Distribution
Inhabits the central mountain chain of New Guinea from Weyland Mountains, Huon Peninsula to south-east Papua New Guinea.

Subspecies
None recorded.

Habits
Considered to be a mid-mountain species although it has been recorded near sea level and as high as 2750 metres. Coates (1985) mentions it as being 'always present in flowering eucalypts and silky oaks in Goroka Town'. This species is common locally although generally considered scarce. It is found in small parties and flocks of up to 40 birds congregating at flowering trees. Their flight is swift and direct.

I was fortunate in seeing this species in Bulolo when a young bird visited the aviaries of Peter Clark. It was also present in nearby higher elevations to the west of the township.

Breeding
The cock pursues the hen, vigorously arching his neck and making a 'churring' type call in courtship, often trying to place a foot on the back of the hen to establish if she is receptive to his advances. The red crown feathers as well as the breast feathers are raised so that the streaks seem enhanced. Slow deliberate walking is also often observed. I suspect that the birds lay their two oval eggs in a hollow of a tree. The only record suggests that they will nest in the dry matted foliage of tall Pandanus trees.

Description
Cock
Length: Approximately 19cm. Weight: Approximately 50–60 grams.

Plumage is predominantly dark green with lighter green underparts streaked with dark green. Underwing coverts are green with a yellow band on the underside of the flight feathers. The bright scarlet-red crown and forehead taper into mauve around the eyes, cheeks and throat. Beak is black and feet are brown. Iris is dark brown.

Hen
Similar to the cock. Trevor Buckell of England has bred many of this species and informed me that the mauve region in line with the eye and between the nape and neck indents with the interruption of green in hens. This has proved to be the case on pairs I have observed. It may not prove to be 100% accurate but so far this has been correct. The cock also tends to have a brighter and larger red crown and more blue around the eye, and the mauve colouration appears to be brighter.

Immatures
Duller than the adults. The forehead is a dull plum colour and the rear of the crown is suffused with blue. Cere and peri-orbital ring are lighter coloured than on the adults. Beak is brown. Iris is grey-brown.

Voice
Quiet hisses with an occasional screech.

Aviculture

The Goldie's Lorikeet is rare in Australian aviaries, but well established in Europe, the UK, South Africa and the USA. This charming, small lorikeet tends to be less confiding than other species.

Goldie's Lorikeets are said to enjoy livefood in the form of mealworms and pupae. I have tried various lories and lorikeets on livefood and have never found them interested. If your birds will eat them then I am sure that it is advantageous to provide protein in this manner. However, it is not necessary if the protein levels are sufficient in the wet and dry food mixtures.

Housing

The minimum suspended aviary size should measure 1.8 metres long x 90cm wide x 1.2 metres high. However, as they will fly very fast in this restricted space, an aviary 3 metres long is preferable.

Breeding

A nestbox measuring 15cm square x 30cm deep would prove adequate for this species. Young have been known to breed at one year of age. Clutch size is two eggs and incubation lasts for approximately 23 days. Young fledge at approximately 56 days of age.

Mutations

There are no known mutations in this species.

MOUNT APO LORIKEET

MOUNT APO LORIKEET
Psitteuteles johnstoniae (Goodfellow)
johnstoniae = named after Mrs Johnstone.

In the Wild
Distribution
Habitat extends over Mount Apo, Mount Malindang and surrounding mountains, Mindanao Island, Philippines.

Subspecies
In my opinion, the subspecies *P.j. pistra* is invalid.

Habits
The Mount Apo Lorikeet inhabits montane forests with a restricted habitat which is under threat by land clearance and hydro-electric schemes. Although found in a National Park, this unfortunately does not guarantee the survival of this species in the Philippines. It is found between 1000–1700 metres in altitude on Mount Malindang and up to 2500 metres on Mount Apo. Goodfellow (1906 in Forshaw 1973) observed flocks of up to 30 birds and reported daily altitudinal movements. During the day birds are found foraging for food in flowering trees or shrubs at higher altitudes, descending to lower altitudes to roost.

Breeding
In the months of March and May adults with enlarged gonads were collected. It is presumed that two eggs are laid in a hollow in a tree.

Description
Cock
Length: Approximately 20cm. Weight: Approximately 55 grams.

The upper plumage is green and the underparts are yellow with green on the edge in a scallop-like pattern. Underwing coverts are yellow-green with a yellow band on the underside of the flight feathers. The forehead, throat and cheek areas are dark pink-orange with a maroon band extending from the forehead through the lores over the eye and the nape. Beak is orange and feet are grey. Iris is dark brown.

Hen
Same as the cock, although the intensity of dark pink around the face may prove to be darker in the cock.

Immatures
The peri-orbital ring and cere are pale grey to white. The face is paler pink in colour and lacks the maroon band across the nape. Beak is black. Iris is dark brown.

Voice
Recorded as being a continual 'lish-lish' sound.

Mount Apo Lorikeet chicks at 8 weeks of age.

Aviculture

This species is extremely rare in captivity. Recently, Mr Antonio de Dios of the Birds International Institute in the Philippines and the Philippine government organised the collection of a number of wild specimens. Since then a number have been bred by Mr de Dios. I understand that breeding is now taking place in the USA and Europe. We can only hope that the limited numbers in Australia will increase.

Housing

A suspended aviary measuring at least 1.8 metres long x 90cm wide x 1.2 metres high would be the minimum requirements to house these birds comfortably.

Breeding

A small nestbox measuring 15cm square x 30cm high should suffice to breed these lovely little birds. Being a montane species I have wondered how they would adapt to breeding in a hot climate. If Mr de Dios has success in Manila, Philippines, then surely there should not be a problem with their breeding in warm to hot climates within Australia.

Clutch size is two eggs and incubation lasts for approximately 23 days. Young fledge between 40–47 days of age.

Mutations

There are no known mutations in this species.

Mount Apo Lorikeet chicks, artificially incubated and handreared, at 3 weeks of age.

VARIED LORIKEET

VARIED LORIKEET
Psitteuteles versicolor **(Lear)**
versicolor = of various colours, variegated.

In the Wild
Distribution
Habitat ranges from the Fitzroy River in the Kimberley region of Western Australia through the 'Top End' of the Northern Territory, south to around Elliot and then across to Cape York with records of sightings on the east coast. My most eastern sighting was at Prairie, 40km east of Hughenden in North Queensland. They are certainly generally scarce to the east of this region.

Varied Lorikeets tend to be found inland rather than on the coast. However, at times, they are seen flying over the sea between the northern suburbs of Darwin and the Bathurst-Melville Island group, Northern Territory.

Subspecies
None recorded.

Habits
Varied Lorikeets are found in noisy flocks of between 20–50 birds occasionally associating in much larger flocks.

Due to their migratory habits to seek out flowering areas, they may be observed in large numbers in an area for some weeks and then vanish. They may return next season or in some cases a few seasons later when the area can again support them. Small flocks fly swiftly between flowering trees in noisy chorus, landing in an appropriate tree. Often squabbles occur and the entire flock will join in a 'free for all'. These minor squabbles do not usually last long and order resumes with each bird moving to individual sprays of flowers.

When flocks are disturbed by a goshawk they tend to congregate into a large flock wheeling above the tree tops with the flock changing directions rapidly to and fro to confuse the predator.

I have found that young Varied Lorikeets tend to be more aggressive than adults in their foraging and usually the adults give way to their persistence.

When an area has flowering trees this does not necessarily mean that flocks of nectar and pollen-eating birds are present. However, when the blossom produces the 'nectar flow' this area will be sought out and converged upon.

During my two year stay in the north of the Northern Territory, I did not see many Varied Lorikeets on the Adelaide River flood plain. However, in mid-October 1986, large flocks were found feeding beside the Arnhem Highway for approximately one week. Although I travelled this road regularly, it was the first time in two years that I discovered this species feeding in this area. This highlights their nomadic nature.

Varied Lorikeets tend to favour dry woodland and eucalypts bordering water courses, particularly so in the Mt Isa and Cloncurry districts.

A possible Pied Varied Lorikeet with nest-mates.

Often flocks are heard long before they are seen when they are traversing country to new feeding grounds. While observing them feeding in flowering trees, I watched a pair with their two offspring hanging upside down (bat-like) for a full 15 minutes. The adults were hanging on both sides of the youngsters and were preening their offspring. A short siesta followed with all birds hanging in this position. I have only recorded this once, so I cannot count this as being a regular habit.

These birds have been recorded feeding on many species of eucalypt, melaleuca and grevillea. I have also seen them feed at the Mt Isa golf course on the Western Bloodwood, *E. terminatis*.

Breeding

Although I have recorded sightings of young Varied Lorikeets in the wild through most of the winter months, unfortunately, I have never located a nest. Generally young are most numerous in the adult flocks between July and November.

Description
Cock

Length: Approximately 19cm. Weight: Approximately 60 grams.

Plumage is predominantly green with a mauve chest and light green underparts streaked with yellow. The forehead, crown and lores area form a scarlet 'cap'. Back of the head and throat are blue-grey streaked with yellow-green. Ear coverts are yellow. Underwing coverts are green. The skin on the cere and the peri-orbital skin are white. Beak is orange and feet are grey. Iris is yellow-orange.

Pair of Varied Lorikeets.

Hen

The intensity of red of the cap is less than on the cock, due to the yellow outer fringe of the red cap feathers. The grey-blue area on the back of the head is darker and the mauve-pink covers a much larger area on the breast of the cock, sometimes extending in two lines down to the inner thigh area. Remember this variation between cocks and hens is not 100% foolproof, but if all these variations are present and the birds are of a similar age it would be unlikely for you to have two of a kind. The yellow ear patch on the cock is usually more intense than on the hen.

Immatures

Duller than the adults particularly in the head and breast regions. The red of the cap is restricted with an orange suffusion going onto the cheeks adjoining the lower mandible. The blue-grey is very much duller, and the yellow ear patch is also duller and less extensive than in the adults. Beak and iris are brown.

It has been discussed that some young leave the nest with a yellowish spot on the back of the head. This is an entirely unreliable way of sexing young Varied Lorikeets and is more often not present at all.

Voice

A shrill screech and a rough grinding type chatter is emitted while feeding, which is unmistakable when trying to locate them in the wild. They also make twittering sounds when they rest, giving the impression of contentment.

Aviculture

Fortunately, Varied Lorikeets are being bred more readily these days due to improved food formulas and avicultural techniques. However, I do not know of any aviculturists who consider themselves to have completely mastered the breeding needs of this species.

My Varied Lorikeets may have eaten some seed, but I cannot confirm this, although they were often seen sitting with the Hooded Parrots at the seed bowl. However, no Varied Lorikeet will stay in good condition on an all-seed diet.

Varied Lorikeets need to be propagated in our aviaries but a note of caution. Do not buy birds from dubious sources, and always strive to obtain guaranteed aviary bred stock.

At present the Varied Lorikeet is not under threat, but then, in the late 1800s people said the same about the Paradise Parakeet.

Housing

I have housed ten Varied Lorikeets in an aviary measuring 6.1 metres long x 3.6 metres wide x 2.2 metres high. They were not the sole occupants of the aviary, which also housed two King Parrots, two Red-capped Lories and two young Hooded Parrots. This was not my ideal, but the birds seemed to 'get on' very well in this collection.

I believe that it is important to obtain as many birds of a species as possible and allow them to select their own partners. I wanted to continue colony breeding this species when I had removed the excess birds from the aviary. However, interference did occur between the pairs and it became necessary to house the pairs separately. I have also found these birds to be extremely tolerant of finches. They will sit near their small associates without harming them. I would suggest that anyone intending to do the same should observe their birds and be satisfied that no trouble will ensue.

My Varied Lorikeets were not fat, but then they had as much exercise as they needed, having not been restricted to small suspended cage systems.

Breeding

Courtship rituals are different from those of other brush-tongued parrot species that I have observed. The pair will bob up and down sitting side by side and then fluff their chests and rump feathers and stretch themselves in unison away from each other with feet placed on the perch. During the stretch both individuals will move their mandibles and arch their necks, possibly emitting a very low noise, which I have not been able to hear. Occasional, more vigorous runs, bobs and jumps occur and are usually performed by the cock. Young begin to bob at approximately two months of age.

Clutch size varies from two to four eggs with the occasional five eggs being recorded in captivity by Barbara O'Brien of Victoria and Stan Sindel of Sydney. Incubation lasts for approximately 22 days. Young fledge by 40 days of age.

Mutations

There are no known established mutations in this species.

LORIUS GENUS

Lorius = possibly of Malay origin signifying parrot.

BLACK-CAPPED LORY
Lorius lory lory

BLACK-CAPPED LORY
Lorius lory lory (Linné)
 This genus was referred to as *Domicella* but this name has been dropped in recent years.
lory = perhaps the same derivation as 'lorius'.
cyanuchen = blue-necked.
erythrothorax = red-breasted.
jobiensis = from the Jobi district on Japen Island.
major = greater.
rubiensis = of the Rubi (district) Geelvink Bay, west Irian.
salvadorii = named after Salvadori.
somu = name of species in local language.
viridicrissalis = pertaining to the green feathers about the crissum, or undertail coverts.

Above: Red-breasted Black-capped Lory *L.l. erythrothorax*.
Left: Black-capped (Jobi) Lory *L.l. jobiensis*.

In the Wild
Distribution
Inhabits New Guinea mainland, western Papuan Islands and various islands in Geelvink Bay, Irian Jaya.

Subspecies
This is a very interesting topic, but a reconciliation of the numerous subspecies is outside the scope of this book. Trevor Buckell (pers. comm.) of the UK has recently completed extensive research on the Black-capped Lory subspecies and is probably the foremost expert on this species in the world. He has extensively studied skins at the Leiden Natural History Museum in Holland as well as the British Natural History Museum.

He now feels that the subspecies can be listed as:

Lorius lory lory	Red underwing coverts
Lorius l. erythrothorax	Red underwing coverts
Lorius l. somu	Red underwing coverts
Lorius l. jobiensis	Blue or black underwing coverts
Lorius l. cyanuchen	Blue or black underwing coverts
Lorius l. salvadorii	Blue or black underwing coverts

L.l. viridicrissalis is inconsistently different from *L.l. salvadorii* and cannot be justified.

L.l. major and *L.l. rubiensis* are unjustified because of the extreme variation in size of the Black-capped Lory within any particular range and also the amount of blue on the breast is so variable with all races.

Habits

L.l. lory is principally a lowland species but has been recorded to 1600 metres in altitude. Above 1000 metres, Diamond (*Avifauna of the East Highlands of New Guinea* 1972) records it as being uncommon. It is found in flowering trees in small family groups, no doubt consisting of pairs with their young of the season. They are certainly not a flocking species. I viewed this species in the mountains west of Bulolo in July 1990 and was amazed that such a colourful bird could be so difficult to observe in the large forest trees of their region. It was considered by Ripley (1937 in Forshaw 1973) to be more solitary on Biak Island, but Thomas Arndt (in *Papageien Magazine*) has observed them again in pairs in recent years. He wrote: 'Three birds were sitting on a palm tree making piercing noises. Once you recognise them they are good to see amongst the palm fronds as their colour is black-red. They were not easy to observe as the later it got the more active they became and would rarely sit still for more than a few seconds. Also, during the following days at different parts of the coast, I noticed that the birds rarely sit on trees other than palms. Before 8am they disappear again.' No doubt they proceed to their feeding grounds. He also notes, that 'the locals do not possess many which often indicates how many there are of a species. The subspecies found here is *L.l. cyanuchen* in which the blue of the nape is not separated from the black cap by a red band. Much of the adjoining islands are being severely deforested and I know of no aviculturist with this subspecies, so I hope that some will fall into the hands of competent individuals so that they can be propagated.'

Black-capped (Salvadori) Lory L.l. salvadorii.

Rand and Gilliard (1967) note: 'It lives in the upper parts of the forest trees where it feeds on flowers. It is only fairly common. Occasionally a number gather in one tree, but usually the species does not travel in flocks and it is more commonly seen, flying or feeding, in pairs.' No doubt, when found in numbers over four individuals, a flock would consist of a couple of family parties congregating solely for a food source, or it could be a group of young birds that have left their family parties. Coates (1985) notes that it is 'quiet and inconspicuous when feeding. Mostly seen in flight when conspicuous and noisy. Flys mostly at treetop level, sometimes higher. Particularly active and noisy in late afternoon.'

Black-capped Lory L.l. salvadorii with yellow and black underwing.

Although there is a difference in size between these subspecies the difference in egg size is marginal.

Black-capped Lory chick at 6 days of age.

Black-capped Lory chick at 18 days of age.

(Photographs on this page by G. Matthews.)

Black-capped Lory chick at 28 days of age.

Black-capped Lory chick at 35 days of age.

Black-capped Lory fledgling at 60 days of age.

Breeding

Ornithologists found cocks with enlarged testes in May and July. 'The male, when displaying, perches in a very upright position and, with wings fully spread and head turned to one side, bobs its whole body up and down (Filewood, pers. comm. in Forshaw 1973). 'A pair was observed excavating a nest hollow in a dead tree on the edge of a garden in the Markham Valley in October' (Watson *et al* 1962 in Coates 1985). We assume that Black-capped Lories lay two eggs in hollows in forest trees.

Description
Cock

Length: Approximately 31cm. Weight: Approximately 200 grams.

Described below are three subspecies of the Black-capped Lory in Australia.

Black-capped Lory *L.l. lory*: The head is red with a black cap. The breast is blue, the blue neckband and nape joining with the breast. The underwing coverts are red with a large yellow band on the underside of the flight feathers.

Red-breasted Lory *L.l. erythrothorax*: Measuring 28–30cm, this subspecies is smaller than the nominate race. The breast colour is red. There is no neckband between the blue-black nape and the lower breast and belly. The underwing colour is also red, with a large yellow band on the underside of the flight feathers.

Salvadori Lory *L.l. salvadorii*: Approximately the same size as the Red-breasted Lory and very similar in appearance, with a red breast and no neckband. However the underwing colour is dark blue to black, with a yellow band on the underside of the flight feathers.

Other subspecies are described below.

L.l. somu: The underwing coverts are red with a large yellow band on the underside of the flight feathers.

L.l. jobiensis, *L.l. cyanuchen* and *L.l. viridicrissalis*: The underwing coverts are blue to mainly black. Iris is red.

Hen
Similar to the cock but often smaller and the black cap is less extensive. These are generalisations and some variation can occur.

Immatures
Duller than the adults and appear smaller on leaving the nest. This is probably due to their being more timid on leaving the nest whereas the adults, particularly the adult cock, are bold and sit in an upright manner. The blue on the breast extends higher than in the adult. Beak is brown. Iris is dark brown.

Voice
Coates (1985) wrote: 'The flight call is an often repeated variable loud musical whistled note, sometimes disyllabic or trisyllabic and often given as pairs of upslurred notes sometimes interspersed with other notes. A variety of loud whistles and piercing squeaks are given when perched.' Black-capped Lories have a large repertoire of calls and in captivity regularly imitate other noises particularly those made by their keepers.

Aviculture
An extremely interesting and wonderful species which was very rare in Australia. Since its importation into this country it is not unusual to see this bird advertised. It is unfortunate that in countries such as Europe and the UK so many hybrids are being bred between different subspecies. There is no excuse for this, as current literature cautions aviculturists to avoid the crossbreeding of subspecies. This is extremely important in aviculture and it must be high on our list of priorities to achieve. In Australia the stock is unfortunately tainted by the crossbreeding. All efforts must be made to try to breed this species true to type.

Housing
As for the Chattering Lory.

Breeding
A nestbox measuring 25cm square x 40cm deep would be sufficient to interest your birds. Tiskens (in *Lori Journaal Internationaal* 1993) used an L-shaped box made with 20mm plywood but he gives no dimensions.

The species is being bred in limited numbers and maintains a high price for a lory in the UK and Europe.

Clutch size is two eggs and incubation lasts for approximately 25 days. Tiskens in Germany records *L.l. jobiensis* as having a 20–21 day incubation period. Young fledge at approximately 60–70 days of age.

Mutations
There are no known mutations in this species.

CHATTERING LORY

CHATTERING LORY
Lorius garrulus garrulus (Linné)
garrulus = chattering.
flavopalliatus = yellow-cloaked.
morotaianus = from the island of Morotai.

In the Wild
Distribution
Northern Moluccan islands, Indonesia.

Subspecies
The nominate race lacks the yellow back and the mantle is completely wine red in colour. It is found on the islands of Halmahera and Weda.

L.g. flavopalliatus is found on Batjan and Obi and has a well-defined yellow patch on the mantle. This bird is often referred to as a Chattering Lory, but in fact should be termed the Yellow-backed Lory or the Yellow-backed Chattering Lory. Many aviculturists in the past did not realise that there are very few of the nominate race being held in captivity.

L.l. morotaianus is said to have a smaller and duller yellow patch on the mantle.

The question arises as to why a large island situated between two smaller groups of islands has a non Yellow-backed form, whereas Yellow-backed forms are found to the north and the south. Does this mean that more research into this area could prove rather interesting in the taxonomy of the Chattering Lory?

Yellow-backed Chattering Lory
L.g. flavopalliatus.

Habits
Rare in the vicinity of human habitation possibly due to the trapping of this species that has occurred. It is usually found in coastal forests. Bishop (1987 in Forshaw 1989) reports its presence on the edges of hills and lowland primary forests. It is also seen in mature woodland up to about 400 metres. It was not recorded in open grassland or dense forest and only occasionally seen in coconut plantations during Bishop's visit to northern Halmahera in October 1987. It certainly seems that it has suffered from habitat destruction and to a lesser extent trapping.

Lendon (1946 in Forshaw 1973) reported inter-pair aggression on Morotai. This species is found in pairs or small family groups and are noisy and conspicuous flying from tree to tree. When feeding they are much quieter and are often difficult to detect. This bird has very good mimic abilities and the local population keep them as pets. A bird I know in the South African collection of Dodds Pringle in the eastern Cape Province was an excellent mimic. It is the best parrot mimic I have ever heard, even better than the African Grey Parrot. It could clearly repeat phrases from both husband and wife and I actually thought the Pringles were speaking until I was informed otherwise. One of my own birds clearly calls my name in my wife's voice causing me many unnecessary trips to the house from my aviaries.

This ability to mimic has unfortunately caused these birds to be trapped and they are a common sight at bird markets in Indonesia and throughout South-East Asia.

Breeding

Lendon, when stationed on Morotai during the war saw fledged young being fed during October and November. He also witnessed a pair investigating a hollow in a tree in June and a young bird was caught in July.

Chattering Lory chicks, at 14 and 16 days of age.

Description
Cock

Length: Approximately 30cm.
Weight: Approximately 200 grams.
Plumage is predominantly bright red with the mantle a deeper red. Wings and thighs are green. Underwing coverts are yellow with a broad pink-red band on the undersides of the primaries. Beak is orange and feet are grey. Iris is red to orange-brown.

Hen

Similar to the cock but usually noticeably smaller in size. The head size of the hen is also narrower and the beak generally smaller.

Immatures

Duller than the adults with green feathering amongst the red of the mantle which is usually lost at about six months of age. Beak is dark brown. Iris is grey.

Voice

It would be wonderful to be able to produce a tape with their varied calls, because for humans it is usually extremely difficult to imitate bird calls. They have a loud nasal 'hoar' type call invariably accompanied by a bob of the head. Loud penetrating whistles

Chattering Lory chick at 20 days of age.

announce their presence or make contact with others, and guttural type calls, in my opinion, signify contentment whilst feeding. There are many other calls which I find not only difficult to describe but also impossible to fathom what the calls could possibly mean.

Yellow-backed Chattering Lory chick approximately 50 days of age.

Aviculture

This species has long been known in captivity but has always been scarce in Australia. In recent years breeding this species has been more successful, probably because we surgically or DNA sex birds and feed them a more nutritious diet as our understanding of their dietary and environmental needs have become better understood.

They are one of the common lories to be found in avicultural collections throughout the world.

Chattering Lories may be noisy, but if you have understanding neighbours they are one of the most interesting and beautiful of birds.

Housing

A suspended aviary measuring a minimum of 3 metres long x 90cm wide x 1.2 metres high is sufficient, however I prefer to allow more room. Once established, they become very tame and will sample food out of your hand or come to the wire.

One of their shortcomings is that they like to give a playful, but extremely painful nip. A tame bird when breeding has very little, if any fear for his keeper, so beware.

Food is best provided from the outside in swivel or rotating feeder bowls, as is usual with most of the *Lorius* species. Using this method can reduce the number of playful nips you have to endure. These feeding bowls are available from manufacturers who advertise regularly in bird magazines.

Breeding

A nestbox placed either vertically or angled at 30–45 degrees will be readily accepted for breeding and roosting. The nestbox size recommended is 25cm square x 60cm long.

Clutch size is two eggs and incubation lasts for approximately 26 days. Young fledge at 70–80 days of age.

Mutations

There has possibly been a Black-eyed Clear mutation of the Yellow-backed Chattering Lory. Whether it is being developed is under question.

Yellow-backed Chattering Lory mutation – possibly a Black-eyed Clear.

YELLOW-BIBBED LORY

YELLOW-BIBBED LORY
Lorius chlorocercus (Gould)
chlorocercus = green-tailed.

In the Wild
Distribution
Eastern Solomon Islands but not found on Bougainville Island.

Subspecies
I have heard from New Zealand breeders, who are of the opinion that there are subspecies based on size coming from various islands within their range. I have no evidence to support this observation.

Yellow-bibbed Lory.

Habits
Found at all altitudes, but said to be more common in the hills on Guadalcanal Island than in its lowland regions. On other islands it is said to frequent coconut groves along the coast. It has been observed to be very active in the crowns of trees searching for food. This would be supported in captivity as this species is by far the most active of all members of the genus.

From reports available, the Yellow-bibbed Lory has a high proportion of seed in its diet as well as nectar, pollen and fruit. G.W. Stevens (*Cage and Aviary Birds* Tabloid 1967–1968) writes that it is easily the most commonly kept pet in the Solomon Islands. 'The usual method of restraining this, and indeed all other parrots is with a double ring of tough coconut shell, rather like a figure eight, with one half around the leg, and the other running along a length of loia vine or rattan – the old-fashioned schoolmaster's cane – with one end terminating in a box shelter on the house verandah, and the other in a nearby tree.'

Breeding
No records known, however, young have been observed in September on Rennell Island. I would assume that the species breeds earlier in the year as do Red-flanked Lorikeets and Eclectus Parrots on Lihir Island, New Ireland Province, Papua New Guinea. Both species breed in July and August. It is assumed that this species lays two eggs in a tree hollow.

Description
Cock
Length: Approximately 28cm. Weight: Approximately 150 grams.

Plumage is predominantly red, wings are green, with a broad band of yellow on the upper breast and a black cap. The black crescents near the yellow bib, when seen in strong sunlight, are in fact dark purple. Underwing coverts are blue with a broad red band on the underside of the flight feathers. Beak is orange and feet are dark grey. Iris varies from dark orange to a brown-orange.

Hen
Same as the cock but generally not as bold in her upright stance.

Immatures
Similar to the adults with a reduced amount of yellow in the bib and in the black crescents joining across the throat. There are often noticeable black patches on the

hindcrown and thighs. Iris is brown. Beak is brownish in colour. Forshaw (1973) states that there are no black markings on the sides of the neck. This is contrary to many young observed in New Zealand and Australia.

Voice

Soft warbles, shrieks, whistles and a variety of sounds which are difficult to describe.

Aviculture

Yellow-bibbed Lory chicks.

Until recently this species was rarely seen in aviculture. It is bred successfully in New Zealand in reasonable numbers. This species is now being exported from the Solomon Islands on a regular basis and is becoming more readily available.

Loro Parque has been breeding this species for a number of years and has been particularly successful with a totally blind hen.

In captivity, it is a very active species, more so than most other *Lorius* members. One unfortunate trait with this bird is its fearlessness. It is, therefore, very prone to biting its keeper. I have often had this species attack my head while I am in the aviary.

I once found one of my Yellow-bibbed Lories unwell and noticed a tick lodged near its ear. I immediately caught the bird and removed the tick. This is the first instance where I had seen a parrot with a tick. The bird was rushed to a veterinarian, treated with anti-tick serum and recovered rapidly. My veterinarian had treated a peacock with a tick previously with similar success. Apparently birds are much less vulnerable to ticks than other domestic animals such as cats and dogs.

It is my experience that Yellow-bibbed Lories will try any food given to them. They are certainly a hardy species.

Housing

A suspended aviary measuring a minimum of 3 metres long x 90cm wide x 1.2 metres high is recommended. When designing an aviary for this quite aggressive species, it is preferable to provide outside swivel or rotating feeder bowls. Nest inspection from the outside would also be advisable.

The birds will benefit from fresh twigs placed inside on a regular basis. Yellow-bibbed Lories seem to enjoy rubbing the sap or pieces of bark through their feathers. This preening process is known as 'anting'.

Breeding

A wooden nestbox measuring 20cm square x 35cm deep is acceptable, however I feel that they probably prefer an inclined box of similar dimensions but a little longer, say 40–50cm.

Cocks are rather aggressive even towards their hens. When I see a pair where the hen is always on the alert and moves away from the cock at every opportunity, I would assume breeding success would be minimal. A hen, in my opinion, should show some fight and then the cock will stop trying to totally dominate her. Even in breeding pairs it is not unusual to see the cock dominate the hen sufficiently for her to retreat. With these facts aside this bird is extremely beautiful and a delightful aviary subject.

Clutch size is two eggs and incubation lasts for approximately 24 days. Young fledge at approximately 70 days of age.

Mutations

There are no known mutations in this species.

PURPLE-NAPED LORY

PURPLE-NAPED LORY
Lorius domicellus (Linné)
domicellus = probably a diminutive of domma which means a little lady or mistress.

In the Wild
Distribution
Ceram and Amboina and also introduced to Buru, Indonesia.

Subspecies
None recorded.

Habits
This extremely beautiful species is found on Ceram between 400–900 metres in altitude and is considered uncommon. This bird needs much closer field study as it is considered vulnerable. There was a sighting in Buru of a bird with a metal ring. This is a common practice in Ceram and Amboina to prevent the birds escaping from pet owners. Deforestation has cleared much of the habitat on Amboina and the bird has only Ceram on which to achieve a viable wild population. It is an unfortunate fact that once a bird's scarcity is promoted in a publication immediate interest in this species is developed. This species has recently been harvested out of Ceram in reasonably high numbers for the avicultural industry. We as aviculturists can only condemn this, unless we harvest sensibly, eg when an area is to be logged. We cannot justify uncontrolled capture.

Breeding
No records are available of breeding in the wild, but I assume that this species nests in tree hollows.

Description
Cock
Length: Approximately 28cm. Weight: Approximately 250 grams.

Plumage is predominantly red with a black cap merging into a purple nape. Wings are green and the upper breast features a yellow band in varying degrees of colour concentration but much less so than that seen on the Yellow-bibbed Lory. Underwing coverts are blue with a broad yellow band on the underside of the flight feathers. Beak and iris are orange-red.

Hen
Similar to the cock. However, in my breeding pair that I owned while living in South Africa, the hen was slightly smaller. The beak on the hen was a darker orange compared to the cock which had a pale orange-yellow beak. This may not prove consistent with all pairs.

Immatures
The yellow bib is more extensive but not as bright as on the adult. The purple on the nape is extensive and is noticeably continued into the black cap. Beak is black changing to orange-yellow. Iris is black-brown.

Voice
The vocabulary of a Purple-naped Lory is without doubt the most varied of any species I have ever kept. I cannot describe most of their calls, but use a tape recording when informing friends. Naturally they screech as all lories do and hiss during courtship.

Aviculture

I would have to say, as difficult as it is to pick and choose between species of lory, that this bird must rank as my favourite.

Although the Purple-naped Lory was considered to be reasonably available in the earlier part of the 20th century, it has never been common. Due to the recent reports of their rarity, a number of specimens have again been made available to aviculture. No more are needed to be taken from the wild and those aviculturists fortunate enough to own them should give them every opportunity to breed.

This species is extremely rare in Australia.

The displays and wonderful repertoire of calls bring back fond memories and I only hope that I will again be fortunate enough to own these birds.

Housing

The minimum recommended size for a suspended aviary is 3 metres long x 90cm wide x 1.2 metres high. My breeding pair in South Africa were housed in a conventional aviary measuring 2.7 metres long x 1.2 metres wide x 1.8 metres high and bred very well in this situation. I would, however, recommend that the aviary be at least 3.6 metres long.

Breeding

Courtship behaviour is typical of other lories and consists of head bobbing, a slow pronounced walk along the perch with an occasional quick flap of the wings, spreading them in such a manner as to show the yellow underside of the flight feathers. Bobbing of the head sometimes follows a figure of eight movement. During these demonstrations the cock hisses and opens his mandibles, flicking his tongue very rapidly. Also, the beak is occasionally clicked together.

My birds bred in a natural hollow measuring approximately 30cm in diameter x 45cm high, attached vertically to the back wall of the aviary. The nest became extremely fouled to an ooze, but the parents resented my interference. It became imperative to clean the nest and once accomplished, the birds quickly returned to their duties. These birds bred in 1975 and subsequent years. In those days few people considered handrearing and my pair consistently reared two clutches per season. I believe, that at the time, they were the most prolific pair in captivity in the world.

Clutch size is two eggs and incubation lasts for approximately 24 days. Young fledge at approximately 63 days of age.

Mutations

There are no known mutations in this species.

PURPLE-BELLIED LORY

PURPLE-BELLIED LORY
Lorius h. hypoinochrous (G.R. Gray)
hypoinochrous = wine coloured beneath.
devittatus = not striped, without a band.
rosselianus = from Rossel Island.

Pair of Purple-bellied Lories.

In the Wild
Distribution
Coastal south-eastern Papua New Guinea west to Cape Rodney on the south and to the southern side of the Huon Gulf on the north. Also found in New Britain, New Ireland, Louisiade and D'Entrecasteaux Archipelagoes, The Duke of York group and other offshore islands.

I am sceptical that the species occur on Tabar and Lihir Islands, as I spent three weeks on Lihir and no specimens were observed. I referred the locals to a colour plate of this species, particularly pointing out the presence of a black cap, and they confirmed that the bird was unknown on Lihir. Some of the locals are good observers in distinguishing bird features. Once you establish certain recognisable features on a bird, they will deliberate amongst themselves and give an accurate reply to your questions. Not one local of the many I asked had ever seen this bird on Lihir. In addition, Ian Burrows (*Report for Lihir Gold Mine*) did not record it on his stay on Lihir during his study of the megapodes.

Subspecies
The nominate race is confined to Misima and Tagula in the Louisiade Archipelago. Its upper breast is lighter coloured than the upper abdomen.

The subspecies *L.h. rosselianus* inhabits Rossel Island in the same archipelago. Its breast and upper abdomen are said to have an even red colour.

The subspecies *L.h. devittatus* is found on mainland New Guinea and all other areas of this species' distribution. This subspecies is said to have no black margins to the greater underwing coverts.

In Coates (1985), a photograph shows a Purple-bellied Lory in flight in the Trobriand Islands. The accompanying caption notes that this bird has black margins to the underwing coverts. Trevor Buckell noticed this and asked me to question Coates, who confirmed that the photograph caption was correct.

The light on the underside of the wing, when the photograph was taken, indicates that it could not be a shadow. This then throws some doubt on the validity of at least one subspecies. More work is obviously necessary on the species.

Habits
The species was said to be extremely abundant in the coconut plantations in the Rabaul region, so much so as to be considered a pest. The government official relating this information even suggested that it may be necessary to control the species. I hope this never becomes a reality, because even though some flowers would be damaged during their foraging, they must still be considered a primary source for pollination of flowers.

The species is common and locally abundant throughout its distribution in primary forests, their margins, tall secondary growth and partly cleared areas. It is said to be found between the coastal lowlands and the foothills to about 750 metres in altitude.

Forshaw (1973) notes that Bell found the species to be 'very plentiful on the Trobriand Islands, particularly around remnants of tall vegetation and in October 1967, on Goodenough Island, in the D'Entrecasteaux Archipelago, he found it to be extremely abundant in secondary growth and gardens, and especially in the flowering coconut palms'.

This very noisy species is conspicuous as it actively searches for pollen, nectar, flowers and fruit. Coates (1985) notes: 'Distinctive in flight with rapid shallow whirring wingbeats, wings appear stubby and rounded.'

Breeding

Breeding takes place mid-year and probably at other times. Young taken from the nest were being handfed by cooks employed at the mine on Misima Island in July. No doubt the species utilises hollows in trees where it lays two eggs.

Courtship consisting of head bobbing, with neck stretching has also been recorded.

Description
Cock

Length: Approximately 26cm. Weight: Approximately 250 grams.

Plumage is predominantly red with black forehead, crown and nape. Abdomen and thighs are purple. Wings are green. Underwing coverts are red, outer margins of which are black. The prominent white cere distinguishes it from all other black cap species of lory. Beak is orange and feet are dark grey. Iris is brown-orange.

Hen

I would assume it to be slightly smaller in build than the cock. As I have only seen four birds in one collection and one in another which had not been sexed, I have no proof of my assumption. However, in viewing the four birds which were housed in one aviary I think this may prove the case on a general basis.

Immatures

Similar to the adults but the bill is brownish.

Voice

Another wonderful species with a variety of unusual calls and screeches.

Aviculture

Although this species is very seldom kept, two birds were known in the Kelling Park collection in England a number of years ago. I saw my first specimen at Peter Clark's aviaries, in Bulolo, Papua New Guinea, where it was kept with several local Black-capped Lories and was friendly with one. The bird was often observed in close contact with its 'mate' and was noticeably larger in build than the other birds in the aviary. It was fed on a pawpaw diet at that time. Special mixtures were difficult to supply to the birds due to the difficulty in obtaining the ingredients in Papua New Guinea. However, all birds appeared fit and healthy.

This species is rare in Australia and we can only hope that they become more available in the future. These lories are wonderful and in my opinion is the only lory that rivals the Purple-naped Lory as being the most desirable to keep.

Housing

Anne Love, when living in Port Moresby, kept four birds together in a large conventional aviary.

One bird could speak and while I was photographing him, he uttered an expletive (abbreviated here as 'AH'). He was certainly to the point. The bird was taught to speak by an expatriate Australian who gave the bird to Anne when he left Papua New Guinea.

Breeding

Unknown in captivity but no doubt similar requirements will induce them to breed as for other members of the genus.

Mutations

There are no known mutations in this species.

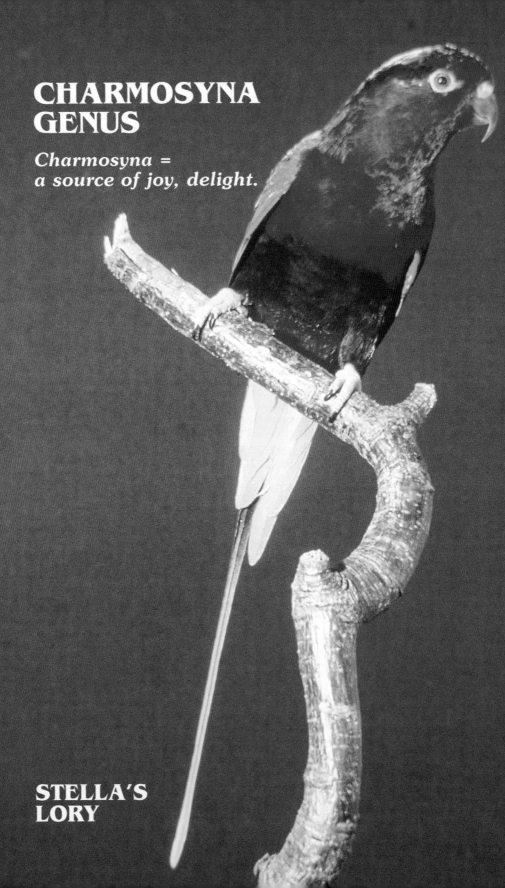

STELLA'S LORY

Charmosyna papou goliathina (Rothshild and Hartert)
papou = *of Papua.*
goliathina = *of Mt Goliath.*
stellae = *presumably named after a lady called Stella.*
wahnesi = *named after the collector/explorer, Wahne.*

In the Wild
Distribution
Central ranges of Papua New Guinea and outlying mountain chains, ie Saruwaged Range, Huon Peninsula.

Subspecies
In captivity the vast majority are the subspecies *C.p. goliathina*, however all birds are referred to as Stella's Lories.

The nominate subspecies is confined to the mountains of Vogelkop, Irian Jaya. We are still not sure if a melanistic phase occurs.

C.p. goliathina is found in the central mountain chain from Geelvink Bay to the Eastern Highlands Province where the subspecies *C.p. stellae* continues the distribution to the east of Papua New Guinea. Both races have melanistic phases. Both have red uppertail coverts, however these are green-edged in *C.p. goliathina*.

I wonder just how consistent these differences are and feel a little sceptical about the validity of separating both races.

C.p. wahnesi has a wide yellow band across the breast and the abdominal patch is said to be washed with green. This subspecies is separated from the central ranges by the Markham River Valley, a wide flood plain, and is found in the mountains of the Huon Peninsula. It is said to have a melanistic phase.

Above: Stella's Lory Melanistic Phase – cock bird.
Below: Bulolo (Papua New Guinea) – habitat of various mountain species.

In *C.p. papou*, the nominate race, of which very few have been known in captivity, the black patch on the head is smaller and restricted to the centre of the crown with very little blue streaking apparent. Yellow patches are found on the sides of the upper breast and the thighs. The blue of the uppertail coverts is more extensive than in *C.p. goliathina* and *C.p. stellae*.

Of all the lories I have observed, this species has the longest tongue, which it often flicks out to clean the sides of its beak. One owner informed me that the bird could reach its eye with its tongue, but I am unable to verify this.

Habits
An inhabitant of forests between 1200–3500 metres in altitude at the extremes. It is considered reasonably common above 2000 metres. I have been fortunate to see this species in both colour phases in the wild, with the red phase birds predominating in small flocks of up to six individuals. They are known to feed on pollen, nectar, fruit, flowers and buds as well as small insects and seed. This is an extremely active species and an incredible sight to see flying between the crowns of trees. The long streamer tail is most noticeable. It has been noted by other observers that the melanistic phase is more common than the red phase in some areas. This species is extremely dexterous among branches, hopping and running between flowers.

Breeding
It would not surprise me to find that the Stella's Lory prefers to excavate its own nest chamber in epiphyte root systems much as the Red-flanked Lorikeet does on Lihir Island. This is not to say that a hollow log would never be used. Pratt (Forshaw 1989) recalled that 'an adult, seen crawling over and under a large clump of epiphytes on the limb of a canopy tree, was thought to be searching for a nest site'.

This occurred in March, at Mt Mengam, Adelbert Range, Papua New Guinea. Hens in breeding condition were collected in October and November. Coates (1985) relates that 'a young bird was obtained at Mafulu, Central Province in November and a fledgling was seen by Hadden at Tari, Southern Highlands Province, probably in October'.

Description
Cock
Length: Approximately 40cm. Weight: Approximately 100 grams.

There are two colour phases – red and melanistic:
Red Phase

Head is predominantly red with dark blue shading to black on the crown. Wings and mantle are green. Lower abdomen is black and the rump is blue. The long tail (approximately 20cm) is green tipped with yellow. Beak is red and feet are pink.

Melanistic

The red is replaced with black. Rump and undertail coverts are red. Beak is red and feet are pink.

Hen
Red Phase

Distinguished from the cock as the sides of the rump and lower back are buttercup yellow instead of red. Underwing coverts are red. Iris is orange-yellow.

Hen
Melanistic Phase

Green replaces the buttercup yellow on the lower back and sides of the rump.

Immatures
Young cocks can be identified in the nest by the red on the lower back and rump area.

W PECKOVER

*Pair of Stella's Lories –
Melanistic cock left and Red hen.*

Duller than the adults with black edging to the red feathers in upper body regions. The tail is shorter. The beak and legs are brownish tending toward orange. Iris is brown.

Voice

Described by Forshaw (1989) as 'a soft, mellow screech, a soft 'cheep...cheep' when hopping about in the tree tops, and while preening or resting a prolonged nasal 'taa...aan'' and by Coates (1985) as a single upslurred grating 'queeea!' when in flight, and a quiet nasal 'wnnaah' when feeding.

Above: Parent reared chicks, six days of age.
Below: Melanistic cock, 29 days of age.

Aviculture

It has been bred on many occasions in captivity, but I would advocate careful management of our captive resources to be assured of a continued supply for aviculture.

Australia has very few of these birds and it is extremely important for aviculturists to obtain new blood through our quarantine facilities. This is the case for many of our exotic avicultural subjects, but we are still hampered by the incredible amount of legislation involved in importation.

Birds are thought to reach 20 years of age in captivity. This is a considerable age for such a species.

One of those wonderful species, which when first seen, is sought after to include in your collection.

Housing

This species should be housed in a suspended cage with minimum measurements of 3 metres long x 90cm wide x 1.2 metres high, but requires a spacious planted aviary to show itself off to its absolute perfection. These birds are extremely active and deserve an adequate area in which to live and breed.

Two young hens, a few days after leaving the nest. Note the short tail. Most hens are identifiable by the yellow back as soon as they leave the nest.

Breeding

A nestbox measuring 20cm square x 45cm long at an inclined angle is adequate.

Mutations

The Melanistic (mutant) phase is recessive to the Red (wildtype) bird in its mode of inheritance. Generally, Red phase birds will breed true to type when paired with a similar partner. Red phase birds can be bred from a Melanistic pair that are split for Red in the cock or hen.

GLOSSOPSITTA GENUS

Glossopsitta = tongue and parrot.

MUSK LORIKEET

MUSK LORIKEET
Glossopsitta concinna (Shaw)
concinna = neat, elegant.

In the Wild
Distribution
Inhabits Cape Upstart in Central Queensland down the east coast to south-eastern South Australia and Tasmania. It is certainly uncommon in Queensland being mainly a temperate climate bird.

Subspecies
None are recognised.

Habits
A very common species in Adelaide, South Australia often seen flying down tree-lined streets of the city.

Musk Lorikeets can be readily heard feeding among blossoms because they are a noisy species. They keep up a constant chatter during their search for food. The predominantly green plumage camouflages them well amongst the foliage and they are often difficult to locate until a movement is detected.

Musk Lorikeets feed on pollen and nectar, and have been recorded feeding on bottlebrush (*Callistemon*), grevilleas and eucalypts. Crop contents have indicated small caterpillars may form part of their diet. Their flight is extremely swift and direct.

Breeding
The breeding season takes place usually from August to January and nests are found in hollow trees generally in lateral branches.

Courtship consists of the cock arching his neck with the head held over the hen. The head is moved to both sides of the hen's head and the cock's pupils dilate. During this performance he chatters, occasionally hisses, puffs out the yellow feathering on the sides of the breast and has the brown nape feathers raised. The hen utters a begging call during this display. The cock frequently moves away from the hen and then returns to her side with an exaggerated hop. The hen arches her back and bows her head to the level of the perch to receive the cock for copulation. After mating, the pair sit close together and preen each other. Youngsters hatch with long white down which is replaced with the second stage down at approximately 12 days of age. This down is grey and very thick and wool-like.

Description
Cock
Length: Approximately 22cm. Weight: Approximately 60 grams.

Plumage is predominantly green with a red forehead and ear coverts. Crown is blue and nape and mantle are brown. Varying dispersal of yellow appears on the sides of the breast. Underwing coverts are green. Beak is coral and feet are grey. Iris is orange.

Hen
Similar to the cock but I have found that most cocks tend to be bluer on the crown

than the hen. The red on the forehead and ear coverts is usually broader. The yellow on the sides of the breast is less extensive and the brown on the nape not as dark as on the cock. Please note that these are my observations and pairs should be confirmed by surgical or DNA sexing.

Immatures

Red areas tend toward orange, there is less blue on the head and the general colouration is duller. The beak is black, however towards the tip there is some light orange which varies in intensity.

Voice

Typical lorikeet screech, but also a warbling type call, which in my opinion is one of contentment.

Aviculture
Housing

Musk Lorikeets require a suspended aviary measuring 2.7 metres long x 90cm wide x 1.2 metres high. Only when seen in a large aviary is their speed in flight appreciated.

Breeding

A nestbox with the internal dimensions measuring 24cm long x 14cm wide x 18cm high with a hollow spout as an entrance proved most adequate for my most prolific pair. My parent birds

Above: Musk Lorikeet chick at 32 days of age. The grey down is very thick and wool-like, possibly due to the colder climate of its habitat.
Right: Musk Lorikeet – Greygreen mutation.

seemed to accept nest inspection, but once they became used to me, they became more aggressive and seemed to resent my interference. However, once the nestbox was cleaned or replaced after inspection, all went back to normal.

I have found Musk Lorikeets to be extremely dedicated parents. They are seasonal breeders, however, a good pair will produce three clutches per season. This is not a burden on the parents provided a good healthy diet is provided.

Mutations

The Greygreen mutation is established. This mutation was originally misnamed 'Olive'.

LITTLE LORIKEET

LITTLE LORIKEET
Glossopsitta pusilla (Shaw)
pusilla = very small.

In the Wild
Distribution
East coast of Australia from the Cairns region, North Queensland through to southeast South Australia and Tasmania.

Subspecies
None recorded.

Habits
A flocking species of small lorikeets, which are often heard before being seen. They are difficult to locate in the upper branches of a blossom-laden tree. I agree with Forshaw, that this species is a canopy feeder and certainly have never seen it in close proximity to the ground except when some wild birds sat in fruit trees next to my captive pairs many years ago in Canberra, Australia.

In those days it was not illegal to capture the birds, and it was a relatively simple matter to direct a fine spray of water onto them with the garden hose. Once the bird was soaked, it would try to clamber away but would invariably fall to the ground making it easy to catch.

C SLANEY

Nest of Little Lorikeet chicks, 32 days of age.

At flowering trees I have witnessed reasonably large flocks, numbering a hundred or so birds greedily feeding among the blossoms. They are usually encountered in smaller flocks of 10–15 birds and in the breeding season, it is not unusual to see family parties foraging together. They have been recorded feeding on the blossoms of eucalypt species, melaleuca and even closer to the ground in *Xanthorrhoea* spp. Loranthus berries and *Euobotrya japonica* fruit have also been recorded in their diet. Their flight is very swift and direct and birds often dart from feeding trees when disturbed by a predator or when alarmed.

Breeding
Their breeding season occurs from May through to January, the birds preferring hollow limbs of trees for nesting. However, birds in the southern part of their range would rarely start before August and September.

The cock courts the hen with an overbearing stance, strutting with chest puffed out. He then bobs his head up and down close to the hen making a low grating sound and often pushing against her as if to force himself on top of her. All the while the chest and head feathers are ruffled with the tail occasionally being fanned out showing the orange-red marking at the base of the lateral tail feathers.

Young hatch with thin white down which is replaced by grey down at about 10 days of age. Young are capable of breeding at 12 months of age.

Description
Cock
Length: Approximately 15cm. Weight: Approximately 45 grams.

Body and wings are predominantly mid-green with red on the forehead, lores, cheeks and throat. Upper mantle and nape are light brown. Underwing coverts are green. Beak is black and feet are pink. Iris is orange.

Hen
Generally less red in the facial mask and not as dark as the cock. The brown nape is also duller.

Immatures
Duller than the adults. Beak and iris are dark brown.

Voice
While flying between trees the birds make a 'tsitt-tsitt' sound, repeated rapidly. The sound makes you aware of their presence. Also known to emit a high-pitched screech.

Aviculture
Due to the increased interest in lorikeets in Australia of recent years, this bird is readily seen in captivity and most are bred on a regular basis. With the advent of more nutritious food recipes, they are easy to cater for and willing breeders.

Housing
A suspended aviary measuring 2.7 metres long x 90cm wide x 1.2 metres high would be acceptable, but this species can be attractively housed in a planted aviary with other inmates such as finches and *Neophema* species. They are bossy, but in my experience, will never persecute other small species, if there is no overcrowding in the aviary.

Breeding
A small Budgerigar type nestbox with a hollow spout as an entrance proves appealing to this species. Other boxes or hollow logs are also accepted.

Mutations
There are no known mutations in this species.

PURPLE-CROWNED LORIKEET

PURPLE-CROWNED LORIKEET
Glossopsitta porphyrocephala **(Dietrichsen)**
Porphyrocephala = purple-headed.

In the Wild
Distribution
South-western Western Australia and from Eden, New South Wales across southern Victoria to southern South Australia, being separated from the western birds by the Nullarbor Plain.

Subspecies
None recorded.

Habits
It is not unusual to find Purple-crowned Lorikeets flying amongst the trees and shrubs in the car park at Melbourne Airport. This species is relatively common throughout its range. Small groups are the norm but large flocks of hundreds can congregate in areas of flowering eucalyptus. Often seen in association with other lorikeets particularly Musk Lorikeets, where the larger species have right of way.

Sindel (1986) records a concentration of thousands of birds at Port Pirie, South Australia related to him by Ross Hogben. 'They were hanging from the blossoms of every flowering tree and scrub throughout the city. Children were plucking them from low blossoms, where some appeared to be intoxicated or perhaps weakened by hunger due to the large concentration sharing the food supply. Many injured birds were brought to Ross after flying into windows or wires. Unfortunately, many more were caught after going into poultry sheds and other enclosures in search of water. Many people who caught these birds thought they would make a pretty pet but did not have the slightest idea of their dietary requirements. Then, as suddenly as they arrived, they departed.'

Pair of Purple-crowned Lorikeets.

I have seen them feeding low down in flowering shrubs at Victor Harbour, South Australia and they certainly were more interested in foraging than my close proximity. Purple-crowned Lorikeets have been reported feeding on eucalyptus and *Melaleuca* spp. as well as boobialla *Myoporum insulare*. It has also been recorded as being an orchard pest.

Breeding
Breeding takes place from August through to December. This species is often recorded as breeding in a loose flock situation. A number of pairs may occupy nest hollows in a single tree with adjoining trees also supporting breeding pairs. Three to four eggs are laid in hollow limbs after the nest chamber has been prepared by both adults. They are said to prefer small entrance holes which I can support from my observations of birds in captivity.

I have observed the cock's courtship display. The cock moves around the hen rapidly with arched neck and beak pulled back into the breast. The purple feathers of the crown

Young Purple-crowned Lorikeets in nestbox.

are ruffled forward and the breast feathers are also raised. If you were close enough, I am sure you could hear the cock hissing through the open bill during his excitement. The hen watches the cock intently, moving away from him as he advances, until she is ready. He then moves next to his mate and rubs his forehead against her neck and head very gently. He then proceeds to mount her as she arches over with raised rump.

The cock probably assists in incubation as the clutch is large for a small bird. However, this assistance is not necessary for success, as he will also sit outside or at the entrance to the nest.

Young hatch with light grey to white down which is replaced from about ten days by thicker, shorter grey down.

Description
Cock
Length: Approximately 15cm. Weight: Approximately 45–50 grams.

Plumage is predominantly light green with a dark purple crown. The area in front of the eyes is orange to orange-red with a lighter shade on the ear coverts. Breast is pale blue and the bend of the wing is a darker, more turquoise shade of blue. Underwing coverts are red and blue leading towards the tip. Beak is black and feet are pink. Iris is dark brown.

Hen
Similar to the cock and difficult to sex. The hen usually has a smaller purple crown and the blue of the body is generally paler.

Immatures
Duller than the adults with little or no purple on the crown.

Voice
A raspy chatter very difficult to describe and not unlike the Red-flanked Lorikeet of New Guinea. Also produces a sharp screech when frightened or when warning other members of a flock.

Aviculture
Most overseas visitors to my aviaries would vote the Purple-crowned Lorikeet as their first choice of Australian lorikeets if the species could be exported. This export obviously does not occur, but it shows the immediate appeal of such a wonderful little creature.

They are reliable breeders and certainly one of my favourite species.

Housing
Purple-crowned Lorikeets will thrive well in a suspended aviary or a conventional planted aviary as described for the Little Lorikeet.

Breeding

These birds have proved consistent and reliable breeders in captivity. A most important consideration is that large clutches can foul the nestbox and they need to be monitored for cleaning. Tragedies can occur in the wet faeces in a nestbox. A Budgerigar type nestbox is readily accepted.

Mutations

The Dilute mutation is established in Australia. It was commonly and incorrectly known as 'Cinnamon', however it is not sex-linked in its mode of inheritance.

Above: Pair of Purple-crowned Lorikeets.
Below: Dilute mutation (left) and Normal Purple-crowned Lorikeets.

NEOPSITTACUS GENUS

Neopsittacus = new and parrot.

MUSSCHENBROEK'S LORIKEET

MUSSCHENBROEK'S LORIKEET
Neopsittacus musschenbroekii (Schlegel)
musschenbroekii = after Musschenbroek.
medius = medium.
major = greater.

In the Wild
Distribution
Central mountains of New Guinea from Vogelkop to south-east Papua New Guinea and the Huon Peninsula.

Subspecies
N.m. medius from the Snow Mountains is probably not sufficiently separable from *N.m. major*, which is said to be larger than the nominate race, with the cheek streaking being brighter and the general plumage duller, particularly the red underparts.

I feel that both *N.m. major* and *N.m. medius* may not be valid, however more study is necessary before they can be dismissed.

Habits
Found as low as 1200 metres and up to 3000 metres in altitude at the extremes, although more commonly encountered between 1600–2500 metres. Said to be common in disturbed areas such as the local gardens in casuarina groves. Usually seen in pairs or in small parties of sometimes up to 50 birds. The Musschenbroek's Lorikeet feeds on pollen, nectar, flowers and buds, fruit, seed and the occasional insect. They are very agile and run dexterously along branches. They have been recorded eating weeds at ground level. Often found in company with other lorikeets, particularly its close relative the Emerald Lorikeet *N. pullicauda* in the higher altitude range. They have a direct and swift flight and seem to be wary of any intruders.

I found them in close proximity to Stella's Lorikeets near Wau although they did not allow a close approach.

Breeding
Little is known, but it has been reported to nest in hollows in trees. I would assume that they may also excavate their own nest chambers in the root ball of epiphytes. A fledgling was taken in mid-November near Lake Habbema, Irian Jaya and a juvenile was obtained in the same area in late August (Coates 1985).

A cock was seen by Beehler (Coates 1985) in pre-copulatory display, south of Wau on December 3, 1975. 'The male bird was seen walking back and forth along a branch, in view of the female. Eventually, the male came to the female and she lowered her head submissively. The male lowered his head in a mimicking posture, and at this point, both birds were frightened off.'

Description
Cock
Length: Approximately 23cm. Weight: Approximately 50 grams.

Upperparts are primarily green and underparts a bright red. Yellow streaking colouration appears on the head and cheeks with a black eye stripe. Forehead is green and the crown and nape are brown. Tail shows yellow and red on the underparts. Underwing coverts and the broad band on the underside of the flight feathers are red. Beak is pale yellow and legs are grey. Iris is red.

Hen
Similar to the cock, but generally duller particularly around the facial area, crown and nape. The brown colouration in this area is darker on the cock.

Immatures
Duller than the adults with blackish margins to the red of the upper chest. A variable yellow band is found on the underside of the flight feathers (Low 1992). The bill is brownish orange.

Voice
A shrill repetitive screech in flight and a disyllabic screech descending in pitch when perched.

Aviculture
I well remember when these birds first became available to South African aviculturists in the late 1970s. A number of pairs were imported and distributed to various breeders. All the birds were found to die within a relatively short time. However, Harry Stephan of the Cape Province had no problem, and in fact bred them within a short time of their acquisition. At that time Harry fed all his lories a wet mix diet. However, every pair also had a bowl of seed, whether they utilised it or not. It was for this reason that the Musschenbroek's Lorikeet survived.

It is an important fact that these little lorikeets require small seed in their diet and when looking at the bird, it will be noticed that they have a more robust beak than other lorikeets. All other breeders who lost their birds did not give their birds seed.

In captivity they are considered to be rather reserved and never confiding. Even pairs that are handreared in most cases seem to revert to a cautious nature.

Housing
As for the Musk Lorikeet.

Breeding
They will breed in a small nestbox measuring 15cm square x 30cm long placed at an inclined or vertical position. To achieve greater breeding success with this extremely interesting bird, dedication is required to develop a more domesticated gene pool.

Mutations
There are no known mutations in this species.

MUTATIONS

Dr Terry Martin BVSc, author of **A Guide to Colour Mutations and Genetics in Parrots,** *kindly contributed his knowledge concerning mutations in lories and lorikeets in the species sections and as follows:*

There has been a recent upsurge in new colours being bred in lorikeets in Australia, particularly in the Rainbow Lorikeet. Whilst there is much contention over the correct name for many of these mutations, they have greatly increased the interest of breeders in this and other lorikeet species.

The Greygreen mutation, which has been introduced into three different birds (the Rainbow, the Red-collared and the Musk Lorikeets) from the original mutation in Scaly-breasted Lorikeets, was originally known as 'Olive'. It is now recognised for what it truly is (a Greygreen mutation), however many breeders still know it by the old name. Add to this situation a new mutation being referred to as the 'true Olive' and confusion is common amongst breeders. Unfortunately this new mutation in the Rainbow Lorikeet, which is co-dominant and produces two colour shades, referred to as 'Jade' and 'Olive', may not be the true Olive either. Research is still needed into this mutation and it will take time to determine exactly what it is.

Another area of naming contention is with the 'Cinnamon' mutations established in Australia for the Rainbow, the Red-collared, the Scaly-breasted and the Purple-crowned Lorikeets. None of these mutations are sex-linked and therefore are all incorrectly named. They contrast distinctly with the true Cinnamon Rainbow Lorikeet bred in Europe and illustrated in this book. At present, the best name for these colours would be Dilute, or Pastel if we adopt the European convention for naming uncertain colours of these shades.

Comparison of the Dilute ('Cinnamon') and the Normal Rainbow Lorikeet.

Few breeders currently realise that the Blue-fronted Rainbow Lorikeet is a melanistic mutation. It is directly equivalent to the Melanistic Stella's Lory. It is combining well with other mutations to produce new colours, as it is one of the few currently available from a different class of mutation to the majority. Dilute, Fallow and Lutino all reduce melanin pigments and do not combine well together. The Melanistic will also combine well with the structural mutations Greygreen and 'Olive'.

The Pied Rainbow Lorikeet is another new mutation that is well established. Like Pied mutations in other species, it adds greatly to the range of colours that can be produced. And like recessive Pied mutations in all species, many split birds carry a few pied feathers indicating the presence of the hidden trait. A dominant Pied has identical appearance in both single factor and double factor, so this is not a true dominant mutation. It is best viewed as a Recessive Pied with 'breakthrough' of splits.

Another mutation being developed is called by the name 'Aqua'. Once again this name appears inappropriate because Aqua is used for Parblue type mutations (also known as Aquamarine), which this does not appear to be. Instead I would currently consider it a structural altering mutation, but study is required to determine exactly how it is functioning. However, whilst I consider the name 'Aqua' to be incorrect, I am hesitant to recommend a new name until we are more certain of the action of this new mutation.

Colour breeding in Rainbow Lorikeets has a bright future and from my perspective is becoming extremely interesting genetically. If a few more new mutations are discovered in Rainbow Lorikeets, we will soon have the 'critical mass' to produce dozens of new combination colours in this species, enough to keep breeders occupied for years to come.

EXPECTATIONS FOR BASIC MUTATIONS

The following are typical genetic outcomes for primary mutations and potential colour combinations.

Primary Mutations

Greygreen – a dominant mutation

SF Greygreen x Normal	=	50% SF Greygreen + 50% Normal
SF Greygreen x SF Greygreen	=	25% Normal + 50% SF Greygreen + 25% DF Greygreen
DF Greygreen x Normal	=	100% SF Greygreen

Melanistic – a recessive mutation

(Dilute, Fallow or Pied could be substituted into these matings.)

Melanistic x Normal	=	100% Normal/Melanistic
Melanistic x Normal/Melanistic	=	50% Normal/Melanistic + 50% Melanistic
Melanistic x Melanistic	=	100% Melanistic
Normal/Melanistic x Normal/Melanistic	=	25% Normal + 50% Normal/Melanistic + 25% Melanistic
Normal/Melanistic x Normal	=	50% Normal/Melanistic + 50% Normal

Lutino – a sex-linked mutation

As this is a sex-linked mutation the sexes involved are important to note.

Lutino cock x Normal hen	=	Normal/Lutino cocks + Lutino hens
Normal cock x Lutino hen	=	Normal/Lutino cocks + Normal hens
Normal/Lutino cock x Normal hen	=	Normal/Lutino + Normal cocks + Normal and Lutino hens
Normal/Lutino cock x Lutino hen	=	Normal/Lutino + Lutino cocks + Normal and Lutino hens
Lutino cock x Lutino hen	=	Lutino cocks and hens

Colour Combinations

Melanistic Greygreen

This combination produces the darkest bird currently possible, an almost black bird. When a Blue becomes established, it will complement this combination even further.

Melanistic x SF Greygreen	=	50% SF Greygreen/Melanistic + 50% Normal/Melanistic
SF Greygreen/Melanistic x Normal	=	25% SF Greygreen/Melanistic + 25% Normal/Melanistic + 25% Normal + 25% SF Greygreen
SF Greygreen/Melanistic x Melanistic	=	25% SF Greygreen/Melanistic + 25% Normal/Melanistic + 25% Melanistic + 25% Melanistic SF Greygreen
SF Greygreen/Melanistic x SF Greygreen/Melanistic	=	6.25% Normal + 12.50% Normal/Melanistic + 6.25% Melanistic + 12.50% SF Greygreen + 25.00% SF Greygreen/Melanistic + 12.50% Melanistic SF Greygreen + 6.25% DF Greygreen + 12.50% DF Greygreen/Melanistic + 6.25% Melanistic DF Greygreen
Melanistic SF Greygreen x Melanistic	=	50% Melanistic + 50% Melanistic SF Greygreen
Melanistic SF Greygreen x Normal	=	50% Normal/Melanistic + 50% Greygreen/Melanistic

Dilute Melanistic

Combining the Dilute gene with the Melanistic gene should produce an attractive colour. Depending on the degree of red in the chest and neck area of the Rainbow Lorikeet strain used, the chest may become similar in colour to the Dilute head. If less red is present in the strain, then the chest colour will approach the back colour of the Dilute.

Dilute x Melanistic	=	100% Normal/Dilute/Melanistic
Normal/Dilute/Melanistic x Normal/Dilute/Melanistic	=	6.25% Normal + 12.50% Normal/Dilute + 6.25% Dilute + 12.50% Normal/Melanistic + 25.00% Normal/Dilute/Melanistic + 12.50% Dilute/Melanistic + 6.25% Melanistic + 12.50% Melanistic/Dilute + 6.25% Dilute Melanistic

Dilute/Melanistic x Melanistic/Dilute	=	25% Normal/Dilute/Melanistic + 25% Dilute/Melanistic + 25% Melanistic/Dilute + 25% Dilute Melanistic
Dilute/Melanistic x Dilute Melanistic	=	50% Dilute/Melanistic + 50% Dilute Melanistic
Melanistic/Dilute x Dilute Melanistic	=	50% Melanistic/Dilute + 50% Dilute Melanistic
Dilute Melanistic x Normal	=	100% Normal/Dilute/Melanistic
Dilute Melanistic x Dilute	=	100% Dilute/Melanistic
Dilute Melanistic x Melanistic	=	100% Melanistic/Dilute
Dilute Melanistic X Dilute Melanistic	=	100% Dilute Melanistic

Dilute Greygreen

The name 'Mustard' has been coined for this colour, although the true Mustard colour requires a true Cinnamon, so we should save the name Mustard for the future.

Dilute x SF Greygreen	=	50% Normal/Dilute + 50% SF Greygreen/Dilute
Dilute x SF Greygreen/Dilute	=	25% Normal/Dilute + 25% Dilute + 25% SF Greygreen/Dilute + 25% Dilute SF Greygreen
Dilute x Dilute SF Greygreen	=	50% Dilute + 50% Dilute SF Greygreen
SF Greygreen/Dilute x Dilute SF Greygreen	=	12.50% Normal/Dilute + 12.50% Dilute + 25.00% SF Greygreen/Dilute + 25.00% Dilute SF Greygreen + 12.50% DF Greygreen/Dilute + 12.50% Dilute DF Greygreen
Normal/Dilute x SF Greygreen/Dilute	=	12.50% Normal + 25.00% Normal/Dilute + 12.50% Dilute + 12.50% SF Greygreen + 25.00% SF Greygreen/Dilute + 12.50% Dilute SF Greygreen
Normal/Dilute x Dilute SF Greygreen	=	25% Normal/Dilute + 25% Dilute + 25% SF Greygreen/Dilute + 25% Dilute SF Greygreen

Dilute Melanistic Greygreen

This triple combination would create an 'all over Mustard' colour. It would be one of the better triple combinations currently possible, along with Pied combinations.

Dilute Melanistic x
Melanistic SF Greygreen
= 50% Melanistic/Dilute
+ 50% Melanistic SF Greygreen/Dilute

Melanistic SF Greygreen/Dilute x
Dilute Melanistic
= 25% Melanistic/Dilute
+ 25% Dilute Melanistic
+ 25% Melanistic SF Greygreen/Dilute
+ 25% Dilute Melanistic SF Greygreen

Dilute Melanistic x Dilute SF Greygreen
= 50% Dilute/Melanistic
+ 50% Dilute SF Greygreen/Melanistic

Dilute SF Greygreen/Melanistic x
Dilute Melanistic
= 25% Dilute/Melanistic
+ 25% Dilute Melanistic
+ 25% Dilute SF Greygreen/Melanistic
+ 25% Dilute Melanistic SF Greygreen

Dilute Melanistic SF Greygreen x
Normal
= 50% Normal/Dilute/Melanistic
+ 50% SF Greygreen/Dilute/Melanistic

SF Greygreen/Dilute/Melanistic x
Dilute Melanistic
= 12.5% Normal/Dilute/Melanistic
+ 12.5% Dilute/Melanistic
+ 12.5% Melanistic/Dilute
+ 12.5% Dilute Melanistic
+ 12.5% SF Greygreen/Dilute/Melanistic
+ 12.5% Dilute SF Greygreen/Melanistic
+ 12.5% Melanistic SF Greygreen/Dilute
+ 12.5% Dilute Melanistic SF Greygreen

Other Combinations

Pied can be easily combined with any of the colours previously mentioned. Simply substitute Pied into the matings for any recessive trait. It always combines most attractively with the darker colours so would go best with the Melanistic and Greygreen mutations.

Fallow, when fully established, will combine well with both the Melanistic and the Greygreen mutation. Simply substitute Fallow for Dilute in the previous matings.

Lutino is currently of little value for combining with other mutations, except for genetic study purposes, as in theory it will mask all currently established mutations. When a Blue or Parblue mutation is established, Lutino will then combine to produce Albino and Creamino colours.

The two new mutations of 'Jade' and 'Aqua' need to be studied further before solid recommendations can be made regarding combination colours with these mutations. What can be said at this stage is that combination with Greygreen would be a poor choice as Greygreen masks all other structural altering mutations and both 'Jade' and 'Aqua' appear to be structural type mutations. Personally, I would like to see the result of the combination of these mutations with Lutino. Whilst I doubt a distinctive colour will be produced, it will help confirm that they are structural mutations. Of the current mutations available Melanistic, Pied, Dilute and Fallow would combine best with these two mutations. However the best results will probably be obtained when a Blue mutation is rediscovered.

TABLE OF PRIMARY MUTATIONS

MUTATION	COMMON NAMES	DISTRIBUTION	AVAILABILITY
Rainbow Lorikeet *Trichoglossus haematodus*			
Blue	Blue	Australia	extinct
Sex-linked Lutino	Lutino (hybrid)	Australia	established
Cinnamon	Cinnamon	Europe	uncertain
Dilute	'Cinnamon'	Australia	common
Bronze Fallow	Fallow	Australia	under development
Grey	Greygreen (hybrid)	Australia	common
Khaki	'Olive'	Australia	under development
Recessive Pied	Recessive Pied	Australia	established
Black-eyed Clear	Yellow	Australia	under development
Melanistic	'Blue-fronted'	Australia	common
Uncertain designation	'Aqua'	Australia	under development
Red-collared Lorikeet *Trychoglossus haematodus rubritorquis*			
Grey	Greygreen (hybrid)	Australia	established
Sex-linked Lutino	Lutino (hybrid)	Australia	established
Dilute	'Cinnamon'	Australia	established
Pied	Pied	Australia	under development
Scaly-breasted Lorikeet *Trichoglossus chlorolepidotus*			
Blue	Blue	Europe	under development
Sex-linked Lutino	Lutino	Australia	established
Dilute	'Cinnamon'	Australia	established
Grey	Greygreen	Australia	established
Musk Lorikeet *Glossopsitta concinna*			
Grey	Greygreen (hybrid)	Australia	common
Purple-crowned Lorikeet *Glossopsitta porphyrocephala*			
Dilute	'Cinnamon'	Australia	established
Stella's Lory *Charmonsyna papou goliathina*			
Melanistic	Melanistic phase	Worldwide	established
Dusky Lory *Pseudeos fuscata*			
Uncertain designation	'Yellow phase'	Worldwide	common
Yellow-backed Chattering Lory *Lorius garrulus flavopalliatus*			
Black-eyed Clear?		Asia	extinct

BIBLIOGRAPHY

Arndt, T., *Papageien*, Harrison, L. (trans) in Parrot Magazine, Parrot Society of New Zealand.
Avifauna of the East Highlands of New Guinea, 1972. Pubs Nuttall Orn. Club. No.12.
Blakers, M., Davies, S.J.J.F., Reilly, P.N., 1985. *The Atlas of Australian Birds*, RAOU Publication, Victoria.
Burrows, I., *Report for Lihir Gold Mine*, unpublished.
Cain, A.J., 1997. *A Revision of Trichoglossus haematodus and of Australian Platycercine Parrots*, 1997, Ibis, pp.432–479.
Cannon, M.J., 2002. *A Guide to Basic Health and Disease in Birds*, ABK Publications, Australia.
Churchill, D.M., Christensen, P., 1970. *Observations on Pollen Harvesting by Brush-tongued Lorikeets*, Journal of Zoology Vol. 18.
Coates, B.J., 1985. *The Birds of Papua New Guinea* Vol. 1, Dove Publications, Brisbane.
Crome, F., Shields, J., 1992. *Parrots and Pigeons of Australia*, Angus and Robertson, Sydney.
Dorge, R., Sibley, G., 2001. *A Guide to Pet and Companion Birds*, ABK Publications, Australia.
Forshaw, J., 1973. *Parrots of the World*, 1st edn, Lansdowne Editions, Melbourne.
Forshaw, J., Cooper, W., 1981. *Australian Parrots*, 2nd edn, Lansdowne Editions, Melbourne.
Forshaw, J., Cooper, W., 1989. *Parrots of the World*, 3rd (revised) edn, Lansdowne Editions, Melbourne.
Lori Journaal Internationaal, Vol 1993 No. 2, Holland.
Lori Journaal Internationaal, Vol 1994 No. 1, Holland.
Low, R., 1977. *Lories and Lorikeets*, Paul Elek Ltd, London.
Low, R., 1986. *Parrots – Their Care and Breeding*, Blandford Press, London.
Low, R., 1987. *Handrearing Parrots and other Birds*, Blandford Press, London.
Low, R., 1992. *Parrots in Aviculture*, Silvio Mattachione and Company, Canada.
Mackay, R., 1971. *Observations for September*, New Guinea Bird Society Newsletter 71/3.
Macwhirter, P., 1987. *Everybird – A Guide to Bird Health*, Inkata Press, Melbourne and Sydney.
Rand & Gilliard, 1967. *Handbook of New Guinea Birds*, Weidenfeld and Nicholson.
Sayers, B.C., 1974. *Aviculture Magazine*, UK.
Shephard, M., 1989. *Aviculture in Australia*, Black Cockatoo Press, Melbourne.
Sindel, S., 1986. *Australian Lorikeets*, Singil Press, Australia.
Smiet, F., 1985. *Notes on the Field Status of Trade in Moluccan Parrots*, Biol. Consrv., No. 34.
Stevens, G.W., approx 1967–1968. *Studying Psittacines in the South Pacific*, Cage and Aviary Bird Tabloid, IPC Magazine Group, London, UK.

RECOMMENDED READING

Australian Birdkeeper Magazine, ABK Publications, Australia. Six issues annually.
Lori Journaal Internationaal, Klein Baal 33, 6685 AC Haalderen, Holland. Four issues annually. This magazine is a must for the lory and lorikeet enthusiast.
Papageien Magazine and *Lexicon of Parrots*, Bruckenfeldstr 30, 75015 Bretten-Rinklingen, Germany. The magazine and Lexicon of Parrots are published in German with the latter also available in English.
A Guide to Incubation and Handraising Parrots, 1998. ABK Publications, Australia.

SPECIES NAME AND WEIGHT TABLE

SCIENTIFIC NAME	DUTCH	GERMAN	ENGLISH	WEIGHT (GRAMS APPROX)
Chalcopsitta a. atra	Zwarte Lori	Schwarzlori	Black Lory	220
C.a. bernsteini	Bernsteinlori	Bernsteins Schwarzlori	Bernstein's Black Lory	200–250
C.a. insignis	Rajahlori	Sammetlori	Rajah Black Lory	200–250
C.d. duivenbodei	Duivenbode's Lori	Braunlori	Duyvenbode's Lory	230
C.s. scintillata	Geelgestreepte Lori	Schimmerlori	Yellow-streaked Lory	200
C. cardinalis	Kardinaallori	Kardinallori	Cardinal Lory	200
Eos cyanogenia	Koningslori	Blauohrlori	Black-winged Lory	160
E.s. squamata	Violetneklori	Bechsteinkapuzenlori	Violet-necked Lory	110
E. reticulata	Blauwgestreepte Lori	Blaustrichellori	Blue-streaked Lory	160
E.h. histrio	Diadeemlori	Diademlori	Red and Blue Lory	160
E.b. bornea	Rode lori	Amboina Rotlori	Red Lory	170
Pseudeos fuscata	Witruglori	Weissburzel Lori	Dusky Lory	155
Trichoglossus ornatus	Ornaatlori	Schmuck Lori	Ornate Lorikeet	95
T.h. haematodus	Groenneklori	Breitbinden Allfarblori	Green-naped Lorikeet	130–150
T.h. mitchellii	Mitchell Lori	Mitchell Allfarblori	Mitchell's Lorikeet	100
T.h. weberi	Weberlori	Webers Allfarblori	Weber's Lorikeet	85
T.h. capistratus	Bloedvleklori	Blauwangen Allfarblori	Edward's Lorikeet	130–150
T.h. rosenbergii	Rosenberglori	Rosenberg Allfarblori	Rosenberg's Lorikeet	130–150
T.h. massena	Massenalori	Massena Allfarblori	Massena's Lorikeet	130–150

T.h. moluccanus	Lori van de blauwe bergen	Gebirgs Allfarblori	Rainbow Lorikeet	130–150
T.h. rubritorquis	Roodneklori	Rotnacken Allfarblori	Red-collared Lorikeet	130–150
T.f. meyeri	Meyerlori	Meyers gelbgrüner Lori	Meyer's Lorikeet	55–60
T. chlorolepidotus	Schubbenlori	Schuppenlori	Scaly-breasted Lorikeet	85
T. euteles	Geelkoplori	Gelbkopflori	Perfect Lorikeet	100–130
Psitteuteles goldiei	Vicoltjeslori	Veilchenlori	Goldie's Lorikeet	50–60
P. johnstoniae	Johnstonlori	Mount Apolori	Mt Apo Lorikeet	55–65
P. versicolor	Bonte Lori	Buntlori	Varied Lorikeet	60–70
Lorius l. lory	Zwartkoplori	Frauenlori	Black-capped Lory	200–250
L.l. erythrothorax	Roodborstlori	Salvadori Frauenlori	Red-breasted Lory	200–230
L.l. salvadorii	Salvadorilori	Meyers Frauenlori	Salvadori Lory	200–230
L.l. jobiensis	Jobilori	Jobi Frauenlori	Jobi Lory	210–250
L. chlorocercus	Groenstaartlori	Grunschwanzlori	Yellow-bibbed Lory	150
L. domicellus	Purperkaplori	Erzlori	Purple-naped Lory	250
L.g. garrulus	Molukkenlori	Gelbmantellori	Chattering Lory	200
L.h. hypoinochrous	Violetstaartlori	Schwarzsteisslori	Purple-bellied Lory	250
Glossopsitta concinna	Muskuslori	Moschuslori	Musk Lorikeet	60
G. pusilla	Dwerglori	Zwergmoschuslori	Little Lorikeet	45
G. porphyrocephala	Purperkroonlori	Blauscheitellori	Purple-crowned Lorikeet	45
Charmosyna papou goliathina	Stellalori	Stella Papualori	Stella's Lory	100
Oreopsittacus arfaki major	Arfaklori	Arfakbergzierlori	Whiskered Lorikeet	22–28
Neopsittacus musschenbroekii	Musschenbroeklori	Gualori	Musschenbroek's Lorikeet	50

Simply the best publications on pet & aviary birds available ...

Six glossy, colourful and informative issues per year. Featuring articles written by top breeders and avian veterinarians from all over the world.

SUBSCRIPTIONS AVAILABLE

The Acclaimed 'A Guide to ...' range

Concise, informative and colourful reading for all bird keepers and aviculturists.

- A Guide to Gouldian Finches
- A Guide to Australian Long and Broad-tailed Parrots and New Zealand Kakarikis
- A Guide to Rosellas and Their Mutations
- A Guide to Eclectus Parrots
- A Guide to Cockatiels and Their Mutations
- A Guide to Pigeons, Doves and Quail
- A Guide to Australian Grassfinches
- A Guide to Neophema and Psephotus Grass Parrots and Their Mutations (Revised Edition)
- A Guide to Asiatic Parrots and Their Mutations (Revised Edition)
- A Guide to Basic Health and Disease in Birds (Revised Edition)
- A Guide to Incubation and Handraising Parrots
- A Guide to Pheasants and Waterfowl
- A Guide to Pet and Companion Birds
- A Guide to Australian White Cockatoos
- A Guide to Zebra Finches
- A Guide to Popular Conures

Soon to be released:
- A Guide to Colour Mutations and Genetics in Parrots

Handbook of
Birds, Cages & Aviaries

For further information or Free Catalogue contact:
ABK Publications
P.O. Box 6288 South Tweed Heads
NSW 2486 Australia

Phone: (07) 5590 7777 Fax: (07) 5590 7130
Email: birdkeeper@birdkeeper.com.au
Website: www.birdkeeper.com.au